T0344592

Solution Synthesis of Inorganic Functional Materials — Films, Nanoparticles, and Nanocomposites

MATERIALS RESEARCH SOCIETY
SYMPOSIUM PROCEEDINGS VOLUME 1547

Solution Synthesis of Inorganic Functional Materials — Films, Nanoparticles, and Nanocomposites

Symposium held April 1–5, 2013, San Francisco, California U.S.A.

EDITORS

Menka Jain
University of Connecticut
Storrs, Connecticut, U.S.A.

Quanxi Jia
Los Alamos National Laboratory
Los Alamos, New Mexico, U.S.A.

Teresa Puig
Institut de Ciencia de Materials de Barcelona, CSIC
Bellaterra, Spain

Hiromitsu Kozuka
Kansai University
Suita-shi, Osaka, Japan

Materials Research Society
Warrendale, Pennsylvania

Shaftesbury Road, Cambridge CB2 8EA, United Kingdom

One Liberty Plaza, 20th Floor, New York, NY 10006, USA

477 Williamstown Road, Port Melbourne, VIC 3207, Australia

314–321, 3rd Floor, Plot 3, Splendor Forum, Jasola District Centre, New Delhi – 110025, India

103 Penang Road, #05–06/07, Visioncrest Commercial, Singapore 238467

Cambridge University Press is part of Cambridge University Press & Assessment, a department of the University of Cambridge.

We share the University's mission to contribute to society through the pursuit of education, learning and research at the highest international levels of excellence.

www.cambridge.org
Information on this title: www.cambridge.org/9781605115245

Materials Research Society
506 Keystone Drive, Warrendale, PA 15086
http://www.mrs.org

First published 2013

A catalogue record for this publication is available from the British Library

CODEN: MRSPDH

ISBN   978-1-605-11524-5   Hardback

CONTENTS

*Invited Paper

FERROELECTRICS AND MULTIFERROICS

MATERIALS FOR ENERGY AND ELECTRONIC DEVICES

NANOMATERIALS AND NANOCOMPOSITES

PREFACE

Symposium M, "Solution Synthesis of Inorganic Functional Materials — Films, Nanoparticles, and Nanocomposites" was held April 1–5th, 2013 at the 2013 MRS Spring Meeting in San Francisco, California.

The symposium was focused on solution synthesis approaches for the growth of a wide range of advanced functional inorganic materials. Recent results were presented on the growth of: (i) highly crystalline functional oxide films; (ii) nanoparticles and nanocrystals; and (iii) nanostructures or nanocomposites by various chemical solution methods. An increased interest in the low-cost and high throughput synthesis of functional and multifunctional inorganic materials indicates the importance of these studies. Gas sensing, photovoltaic, plasmonics, memory devices, spintronics, bio-medical, superconducting, and magnetic-field sensing applications were extensively discussed. The symposium promoted information exchange between worldwide researchers from universities, national labs, and industries.

At this symposium, 225 abstracts were presented and more than 100 attendees attended the sessions. Both oral presentations and poster sessions were held.

The articles in this symposium proceeding volume cover the development of different chemical solution approaches to synthesize inorganic functional materials for enhanced and/or novel functionalities for a variety of applications. The papers in the volume have been divided into four sections: (1) thin film preparation methods, (2) ferroelectrics and multiferroics, (3) materials for energy and electronic devices, and (4) nanomaterials and nanocomposites. These papers convey the breadth of exciting advancements happening in the area of functional materials grown by various solution methods.

Menka Jain
Quanxi Jia
Teresa Puig
Hiromitsu Kozuka

September 2013

ACKNOWLEDGMENTS

The papers published in this volume result from the MRS Spring 2013 Symposium M. We sincerely thank all of the oral and poster presenters who contributed to this proceeding volume. We also thank the reviewers who provided valuable feedback to the editors and authors. We greatly appreciate the MRS publication staff for their constant help and for guiding us smoothly through the submission/review/decision process.

MATERIALS RESEARCH SOCIETY SYMPOSIUM PROCEEDINGS

MATERIALS RESEARCH SOCIETY SYMPOSIUM PROCEEDINGS

Volume 1562E — Emerging Materials and Devices for Future Nonvolatile Memories, 2013, Y. Fujisaki, P. Dimitrakis, D. Chu, D. Worledge, ISBN 978-1-60511-539-9

Volume 1563E — Phase-Change Materials for Memory, Reconfigurable Electronics, and Cognitive Applications, 2013, R. Calarco, P. Fons, B.J. Kooi, M. Salinga, ISBN 978-1-60511-540-5

Volume 1564E — Single-Dopant Semiconductor Optoelectronics, 2014, M.E. Flatté, D.D. Awschalom, P.M. Koenraad, ISBN 978-1-60511-541-2

Volume 1565E — Materials for High-Performance Photonics II, 2013, T.M. Cooper, S.R. Flom, M. Bockstaller, C. Lopes, ISBN 978-1-60511-542-9

Volume 1566E — Resonant Optics in Metallic and Dielectric Structures—Fundamentals and Applications, 2013, L. Cao, N. Engheta, J. Munday, S. Zhang, ISBN 978-1-60511-543-6

Volume 1567E — Fundamental Processes in Organic Electronics, 2013, A.J. Moule, ISBN 978-1-60511-544-3

Volume 1568E — Charge and Spin Transport in Organic Semiconductor Materials, 2013, H. Sirringhaus, J. Takeya, A. Facchetti, M. Wohlgenannt, ISBN 978-1-60511-545-0

Volume 1569 — Advanced Materials for Biological and Biomedical Applications, 2013, M. Oyen, A. Lendlein, W.T. Pennington, L. Stanciu, S. Svenson, ISBN 978-1-60511-546-7

Volume 1570E — Adaptive Soft Matter through Molecular Networks, 2013, R. Ulijn, N. Gianneschi, R. Naik, J. van Esch, ISBN 978-1-60511-547-4

Volume 1571E — Lanthanide Nanomaterials for Imaging, Sensing and Optoelectronics, 2013, H. He, Z-N. Chen, N. Robertson, ISBN 978-1-60511-548-1

Volume 1572E — Bioelectronics—Materials, Interfaces and Applications, 2013, A. Noy, N. Ashkenasy, C.F. Blanford, A. Takshi, ISBN 978-1-60511-549-8

Volume 1574E — Plasma and Low-Energy Ion-Beam-Assisted Processing and Synthesis of Energy-Related Materials, 2013, G. Abrasonis, ISBN 978-1-60511-551-1

Volume 1575E — Materials Applications of Ionic Liquids, 2013, D. Jiang, ISBN 978-1-60511-552-8

Volume 1576E — Nuclear Radiation Detection Materials, 2014, A. Burger, M. Fiederle, L. Franks, D.L. Perry, ISBN 978-1-60511-553-5

Volume 1577E — Oxide Thin Films and Heterostructures for Advanced Information and Energy Technologies, 2013, G. Herranz, H-N. Lee, J. Kreisel, H. Ohta, ISBN 978-1-60511-554-2

Volume 1578E — Titanium Dioxide—Fundamentals and Applications, 2013, A. Selloni , ISBN 978-1-60511-555-9

Volume 1579E — Superconducting Materials—From Basic Science to Deployment, 2013, Q. Li, K. Sato, L. Cooley, B. Holzapfel, ISBN 978-1-60511-556-6

Volume 1580E — Size-Dependent and Coupled Properties of Materials, 2013, B.G. Clark, D. Kiener, G.M. Pharr, A.S. Schneider, ISBN 978-1-60511-557-3

Volume 1581E — Novel Functionality by Reversible Phase Transformation, 2013, R.D. James, S. Fähler, A. Planes, I. Takeuchi, ISBN 978-1-60511-558-0

Volume 1582E — Extreme Environments—A Route to Novel Materials, 2013, A. Goncharov, ISBN 978-1-60511-559-7

Volume 1583E — Materials Education—Toward a Lab-to-Classroom Initiative, 2013, E.M. Campo, C.C. Broadbridge, K. Hollar, C. Constantin, ISBN 978-1-60511-560-3

Prior Materials Research Symposium Proceedings available by contacting Materials Research Society

Thin Film Preparation Methods

Mater. Res. Soc. Symp. Proc. Vol. 1547 © 2013 Materials Research Society
DOI: 10.1557/opl.2013.635

Thin YBa$_2$Cu$_3$O$_{7-\delta}$ patterns by Chemical Solution Processing using Ink-Jet Printing

Jonas Feys[1], Bram Ghekiere[1], Petra Lommens[1], Simon C. Hopkins[2], Pieter Vermeir[1,3], Michael Baecker[4], Bartek A. Glowacki[2,5,6], Isabel Van Driessche[1]

[1] SCRiPTS, Ghent University, Ghent, Belgium
[2] ASCG, Department of Materials Science and Metallurgy, University of Cambridge, Cambridge CB2 3QZ, UK
[3] Department of Industrial Sciences; University College Ghent; Ghent; Belgium
[4] Deutsche Nanoschicht GmbH, Rheinbach, Germany
[5] Department of Physics and Energy, University of Limerick, Ireland
[6] Institute of Power Engineering, ul Augustowka 6, 02-981 Warsaw, Poland

ABSTRACT

In this paper, we present ink-jet printing as an attractive alternative to lithography and etching methods for the development of multi-filamentary YBa$_2$Cu$_3$O$_{7-\delta}$ coated conductors. Our research is mainly focused on the study of the influence of rheological parameters on the printability of water-based inks in order to produce superconducting patterns on SrTiO$_3$ and CeO$_2$-La$_2$Zr$_2$O$_7$-Ni5at%W substrates. An aqueous YBCO precursor ink with a total metal ion concentration of 1.1 mol/L with a viscosity of 6.79 mPa s and a surface tension of 67.9 mN/m is developed. Its printing behavior using several ink-jet printing devices is verified using a camera with strobed illumination to quantify droplet velocity and volume. After optimization of the deposition parameters, YBCO tracks with different dimensions could be printed on both types of substrates. Their shape and dimensions were determined using optical microscopy and non-contact profilometry, showing 100-200 nm thick and 40-200 µm wide tracks. Finally, resistivity measurements were performed on the widest tracks on SrTiO$_3$ showing a clear drop in the resistivity starting from 88.6 K with a ΔT_c of 1.4 K.

INTRODUCTION

In recent years, several companies have developed technologies for the processing of long lengths of YBa$_2$Cu$_3$O$_{7-\delta}$ (YBCO) coated conductors [1-5]. One common coated conductor design consists of a well textured Ni-5at%W substrate, covered by a buffer layer architecture topped with a continuous superconducting film. These conductors can be wound as coils and, when used in motors and generators instead of copper wires, they can reduce the total size and weight of the device because of their ability to carry high currents without resistance. Yet, these coated conductors produce large losses when used with alternating current (AC) or fields [6-9]. The high aspect ratio of the coated conductor geometry leads to high hysteresis losses proportional to the width of the superconducting film. These losses can be effectively reduced by dividing the superconducting film into narrow filaments, with the net hysteresis losses ideally inversely proportional to the number of filaments. In practice, the total AC losses of a striated superconducting tape contain other contributions which would also need to be considered in a coated conductor optimized for AC applications. There will always be eddy current losses associated with substrate and stabilization layers, which can be reduced by maximizing the substrate resistance and by also striating the stabilization layer. These methods, and avoiding interfilamentary bridging, also reduce coupling currents; but twisting the conductor, or ideally fully transposing the filaments, is required to minimize coupling losses [10-18]. This filamentary superconducting structure can be accomplished through several techniques which can be divided into two main approaches: with or without material removal. Lithography, wet chemical etching, ion beam etching, laser ablation or slitting of the tape are techniques where superconducting material is removed [7, 8, 12, 19-21]. They have a variety of disadvantages such as high cost, local degradation of the superconducting properties or tape fragmentation. The other class of techniques allows direct patterning of the superconductor through electrodeposition or drop-on-demand (DOD) inkjet printing [22-26]. While ink-jet printing has been used for many years in depositing text or patterns on textiles or paper, it is only recently that there has been a growing interest in using the deposition technique for functional ceramic coatings and structures [27-31]. The biggest advantage of ink-jet printing as a one-step process is its low investment cost, scalability, the more efficient use of materials and the good control of the thickness and pattern of the coating.

In this paper, we show that by combining the ink-jet printing method with completely water-based inks, containing non-hazardous metal salts as precursor materials, it is possible to print multi-filamentary superconducting YBCO patterns. Avoiding the incorporation of F-containing components results in a faster and environmentally more benign production method. By controlling the wetting behavior of the ink on appropriate substrates, it becomes possible to predict pattern dimensions. Therefore, both the droplet diameter after ejection from the printing nozzle and contact angle after contacting the substrate need to be known. The droplet diameter is mainly dependent on the opening of the nozzle, therefore different nozzle orifices were used for producing the ink patterns. The inverse value of the Ohnesorge number of the ink, an important parameter predicting its printability depending on the nozzle dimensions and the rheology of the ink, was verified to assess its printability in the different nozzles. When using piezo-electric ink-jet printing, it was possible to vary YBCO pattern dimensions from 40-200 μm in width and 200-250 nm in height on both buffered $SrTiO_3$ single crystals and buffered Ni-5at%W tape.

EXPERIMENTAL DETAILS

The aqueous precursor solution was prepared by dissolving $Y_2(CO_3)_3.1.9H_2O$ (99.9 %, Sigma Aldrich), $Ba(OH)_2.8H_2O$ (98 %, Janssen) and $Cu(NO_3)_2.2.5H_2O$ (98 %, Alfa Aesar) salts in water and nitrilo-triacetic acid (NTA, 99 %, Alfa Aesar). The addition of triethanolamine (99+ %, Acros Organics) increases the pH to 6-8 and the viscosity to 6.79 mPa s (22 °C, shear rate of 9.10^4 s^{-1}, Haake Technik rheometer) respectively. The total metal ion concentration of the precursor solution was 1.1 mol/L (0.185 mol/L YBCO), as verified by ICP-OES (Spectro, Genesis) [26].

The density of the aqueous solution was obtained using a 10 cm^3 glass pycnometer (Duran). The surface tension of the precursor solution and contact angles on different substrates were measured with a DSA30 instrument (KRÜSS, GmbH, Germany).

The as-received $SrTiO_3$ (STO, CrysTec) and CeO_2 - $La_2Zr_2O_7$ buffered Ni-5at%W (Deutsche Nanoschicht GmbH) substrates were cleaned in isopropanol prior to ink-jet printing in order to remove dust and organics, and dried in air.

The substrates were coated at room temperature using both a single and a multi-nozzle piezoelectric DOD ink-jet system. By applying pulses of a specific electrical waveform to the piezoelectric actuators, the ink chamber is deformed resulting in a pressure wave, which draws ink into the region of the orifice and expels it, forming a drop in flight. A Dimatix materials printer (Fujifilm Dimatix Inc. DMP-2800), with cartridges providing orifice diameters of 9 and 23 μm (~ 1 and 10 picoliter droplets) respectively, was used as the multi-nozzle system; and for the single nozzle system, Microfab nozzles with a 30 and 60 μm orifice diameter (MicroFab Technologies, MJ-ABP-01-30) and an XY positioning system under computer control were selected. Drop formation and fluid behavior was also imaged in-situ during printing, using a camera (Allied Vision Technologies, Stingray F-125B) with a telecentric zoom lens (Moritex, MLZ07545) and strobed collimated LED illumination (bespoke, University of Cambridge). Custom-written software was used to control image acquisition and to quantitatively analyse the drop volume and velocity. The ink was filtered with a 0.2 μm pore size PET filter prior to printing. Printing was performed using a unidirectional raster pattern, with correct drop positioning achieved by computer-controlled synchronization of the movement of the printer nozzle and ink jetting. All samples were dried at 120 °C in air.

Subsequently, the samples on $SrTiO_3$ were heat treated at 20 °C/min from room temperature to 780 °C in a humid 100 ppm O_2/N_2 atmosphere, and a maximum temperature of 760 °C was selected for patterns deposited on buffered tape [26, 32]. After 90 min dwell time, the gas was switched to dry 100 ppm O_2/N_2 for 40 minutes. Afterwards, the temperature was decreased to 400 °C at 10 °C/min. During this cooling stage, a gas switch to pure O_2 took place at 520 °C. After annealing for 60 min, the samples were furnace cooled to room temperature.

The crystallinity of the processed films was verified using X-ray diffraction (Bragg-Brentano configuration, Siemens, D5000; Cu-Kα). The sample morphology was characterized using optical microscopy (Leitz, Laborlux 12 POL S) and SEM (FEI Nova 600 Nanolab Dual-Beam FIB). A cross section of the layers was made using the FIB module coupled with SEM to verify the thickness of the layers. The shape of the multi-filamentary patterns was visualized by using optical profilometry (Veeco, NT9080). The critical temperature of the superconducting patterns was measured by resistivity measurements using a custom-made four-point test device (Keithly).

DISCUSSION

In recent years, a lot of research has been dedicated to the development of low cost, fluorine-free, preferably aqueous precursor solutions for the chemical deposition of YBCO and buffer layers [33-36]. We have recently published that the use of chelating agents such as NTA and triethanolamine to increase the ion solubility can lead to excellent YBCO precursors [26, 37]. While initially the optimization of the ratio of the chelating agents versus metal ion concentration and the pH were mostly done by trial and error and extensive lab work, we have shown later that it is possible to obtain a better insight into the chemistry by using both theoretical speciation distributions and experimental data obtained by electron paramagnetic resonance (EPR) revealing the related coordination chemistry. In our case we found that a NTA : total metal concentration ratio of 0.45 and a pH range of 6 to 7 were the most suitable formulations leading to precursor solutions with a total metal concentration of 1.1 mol/L that are stable for several months.

In order to have a better insight into the ink printability, we need to ensure that the physical properties of the ink fulfill the criteria for ink-jet printing. The pressure wave inside the print head needs to overcome the viscous flow and the surface tension of the ink. If the viscosity is too high, the pressure wave is damped, inhibiting the jetting of an ink droplet. The generation of droplets in a DOD printer is a complex process, and the precise physics and fluid mechanics of the process are the subject of much research. The behavior of inks in the printing system can be quantified by a number of dimensionless groupings of physical constants, i.e. the Reynolds (Re), Weber (We) and Ohnesorge (Oh) numbers:

$$Re = \frac{vl\rho}{\eta}, We = \frac{v^2 l\rho}{\sigma}, Oh = \frac{\sqrt{We}}{Re} = \frac{\eta}{\sqrt{\sigma\rho l}} \tag{1}$$

with σ, ρ, η and v the ink surface tension, density, viscosity and velocity respectively and l the diameter of the orifice of the nozzle. The Reynolds number is a ratio of internal and viscous forces and the Weber number represents the ratio between the internal and surface tension forces. The inverse value of the Ohnesorge number is a characteristic value which is independent of the droplet velocity. It has been observed that Oh^{-1} should be higher than 2 to ensure proper ink jetting, and more recently Reis and Derby proposed a range of $1 < Oh^{-1} < 10$ [29, 38]. The lower limit is a result of the viscous dissipation of the pressure wave and the upper limit indicates the region in which the formation of satellite drops dominates over a single droplet. This upper limit can be relaxed, as long as the satellites merge with the main droplet before impact on the substrate.

In Table I we show the fluid properties determined for the aqueous YBCO ink used in this work. Since different printing systems are used, the inverse Ohnesorge numbers for the different orifice diameters are presented in Table II.

Table I. Fluid properties of the YBCO ink.

Surface tension σ [J m^{-2}]	Density ρ [kg m^{-3}]	Viscosity η [Pa s]
6.79×10^{-2}	1233	6.8 x 10-3

Table II. Different printing systems used with the corresponding inverse Ohnesorge number for the YBCO ink.

Printing system	Nozzle diameter l (x 10^{-6} m)	Oh^{-1}
DMP2800	9	4.04
	23	6.31
Microfab	30	7.37
	60	10.42

In order to examine the droplet formation process, drop visualization was performed using the Microfab (30 and 60 μm) piezoelectric nozzles and in-house written software. An anti-symmetric bipolar waveform was applied to the piezoelectric element with a maximum voltage of 16 V and 22 V for the 30 μm and 60 μm orifices respectively. From figure 1a, one can see the different stages during ink-jet printing. In the first stage, a growing filament is visible, where after some time, the formation of a droplet starts due to the surface tension forces of the ink. The droplet remains attached to the nozzle by an elongated filament. The detachment of the filaments begins at the nozzle, creating a tail behind the ejected droplet. The relatively high surface tension of the

6

aqueous ink ensures the recombination of the tail with the main drop. Also visible in figure 1a are the secondary oscillations of the ink due to the propagation of the wave pulse in the ink chamber. With the software, both droplet velocity and total volume can be calculated as shown in figure 1b. One droplet contains 90 pL of ink.

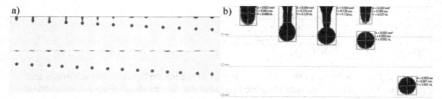

Figure 1. (a) Jetting analysis for the YBCO ink, using a 60 μm Microfab nozzle fired at 22 V with 5 μs time intervals with a final travelled distance of 340 μm and (b) highlighting the important phases during the jetting, images taken at delay times of 35, 65, 75, 85 and 205 μs.

Multi-nozzle printing was performed with the Dimatix materials Printer. This printer has a built-in camera for drop watching which can be used for calculating the trajectory velocity of the droplet as represented in figure 2. The standard waveform consists of a negative voltage peak to suck ink into the ink chamber, a maximum positive voltage to dispense the ink and a segment that allows the nozzle to recover to its original shape without sucking air in the chamber. We found that for the 1 pL and the 10 pL cartridges, the use of a maximum voltage of 8.2 V and 11 V at 33 °C resulted in satellite free droplet formation.

Figure 2. Jetting analysis for the YBCO ink at 11 V and 33 °C, using a 10 pL DMP cartridge.

After finding the best suited jetting parameters, i.e. no satellite formation and a well-defined vertical droplet trajectory, the wetting of the ink onto the different substrates needs to be characterized and optimized. As published before, this can be done by using the pendant drop method for optically determining the surface tension (σ) of the ink and also by measuring the different contact angles (θ_{eqm}) of the ink on the substrates [26]. The surface tension of the substrate was determined by measuring the contact angle for water and diiodomethane. Using these methods, the wetting behavior of the ink on both the single crystal SrTiO₃ (STO) and the buffered Ni-5at%W substrates could be quantified (Table III). The large uncertainty (from 5 – 10 °) in the contact angle values is due to local dewetting effects on the substrate and for further analysis, an upper bound contact angle of 10 ° was used, since the worse value will determine the maximum inter-droplet distance during printing.

Table III. Surface tension and wetting behavior of the YBCO ink and the substrates used.

Liquid / Solid	σ [mN/m]	σ_{polar} [mN/m]	$\sigma_{dispersive}$ [mN/m]	θ_{eqm} [degrees]
STO	77.9	34.6	43.3	5 – 10
Buffered tape [a]	77.8	33.5	45.3	10 – 15
YBCO solution	67.9	41.9	26	/
[a] Ni-5%W tape / LZO / LZO / CeO$_2$				

When combining the knowledge of the wetting behavior of the ink on the different substrates and the droplet dimensions, it now becomes possible to predict the pattern dimensions using equation 2 [29].

$$d_{con} = d_0 \left(\cfrac{8}{\tan\cfrac{\theta_{eqm}}{2}\left(3+\tan^2\cfrac{\theta_{eqm}}{2}\right)} \right)^{1/3} \qquad (2)$$

Where d_{con}, d_0 and θ_{eqm} are the droplet diameter after contact onto the substrate, the droplet diameter in flight and the contact angle at equilibrium. Using this equation, the minimum width (equal to d_{con}) of a pattern can be predicted. In order to increase the thickness of the YBCO filament, several successive coatings were deposited, without an intermediate drying step. In Table 4, we summarized the data for the different nozzles used for printing the YBCO ink onto SrTiO$_3$ substrates. A large difference in the droplet speed in flight (υ) can be observed for Dimatix versus Microfab printing. This is probably due to the differently constructed nozzles and to variations in the wave-form. After calculating the droplet diameter in flight, the droplet diameter after contact was calculated using equation 2 and a contact angle of 10 °. This value was then used for predicting the inter-droplet distance (IDD) that should be used for printing. Although the inter-droplet distance was chosen to have a slightly lower value than d_{con}, the pattern width did not increase when using the 9 and the 22 μm nozzles orifices. For the others, a larger variation was observed in the line width. This can be due do the lower repeatability doing the multiple coatings and due to the local wetting differences. A 5 ° contact angle would result in a spreading of 220 μm starting from a $d_0 = 61$ μm.

Table IV. Summary of the data for printing YBCO ink onto SrTiO$_3$ substrates.

Φ orifice [μm]	volume [pL]	υ [m s^{-1}]	Re	We	d_0 [μm]	d_{con} [μm]	IDD [μm]	line width [μm]	# coatings
9	1	/	/	/	12.4	38	29	**40**	7
22	10	7.86	31.4	24.7	28	87	44	**90**	7
30	22	1.68	9.1	1.5	40	125	100	**160 ± 25**	4
60	90	1.24	13.5	1.7	61	190	140	**190 ± 30**	4

Even when depositing multiple coatings on top of each other, the line width after thermal processing is in good agreement with the predicted value.

Figure 3 shows the digital photographs and optical profilometry images for the different samples. Fine, straight and well-separated filaments exhibiting almost no coffee-ring effect can be observed. Local dewetting effects present sample b and c need to be prevented in the future in order to have a continuous pattern. Breaks in a few filaments would simply reduce the current-carrying cross-sectional area slightly in short samples, but this issue must be addressed for long-length production to be feasible. When taking a closer look at the pattern shape, nice straight edges and a homogeneous thickness distribution are found. The pattern width ranges from 40 to 190 µm as summarized in Table IV. The YBCO tracks have a thickness between 100 and 250 nm after full thermal processing. The distance between the different lines is between 150 and 200 µm. With these distances it becomes possible to divide the YBCO coating into 49, 32, 24 and 24 lines for a 10 x 10 mm single crystal substrate. This is important because it is reported in the literature that the hysteresis losses will be decreased by the inverse of the number of lines ($1/N_a$) in comparison with a completely covered substrate [15, 22]. The currently reported width for patterns is going from 30 µm up to one millimeter when using lithography or etching methods [7, 8, 12, 11, 16, 19, 21, 22, 24]. With our ink-jet printing technique we can obtain a similar lateral resolution in only one processing step. Since the reported YBCO thickness is around 1 µm, modifications to the ink, the deposition parameters and the thermal process will be necessary in order to exceed the 200 nm that we currently obtain.

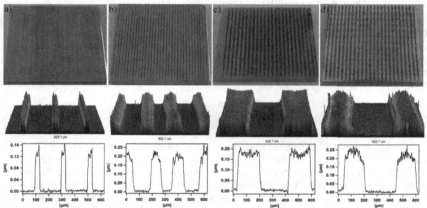

Figure 3. (Top) Optical images, (middle) 3D and (bottom) 2D line scans obtained through optical profilometry for YBCO printed on 10 x 10 mm SrTiO₃ substrates with nozzles with from (a) to (d), a 9, 22, 30 and 60 µm orifice diameter respectively.

The crystallinity of the YBCO patterns after thermal processing was verified using θ-2θ X-ray diffraction (figure 4a). It can be seen that the preferred (001) reflections are dominant on the SrTiO₃ substrate, but also some reflections indicative of the presence of a $BaCuO_x$ phase (*). This phase is often observed on the surface of the YBCO films, deposited by a chemical solution deposition method [26]. The weak reflections appearing at 20.5°, 41.8° and 21.8°, 44.5° and

44.9° are satellites of the large STO substrate peaks due to W-L$_{\alpha1}$ and Cu-K$_\beta$ radiation respectively.

The resistivity measurement obtained for ink-jet printed YBCO patterns, using the 60 μm nozzle, on SrTiO$_3$, is shown in figure 4b. Three lines were connected using silver paste. The measurement was performed in liquid nitrogen and results in a superconducting transition with a T$_{c,onset}$ of 88.6 K and a ΔT$_c$ of 1.4 K. The resistivity does not decrease completely to zero, which may be due to the presence of the BaCuO$_x$ phase.

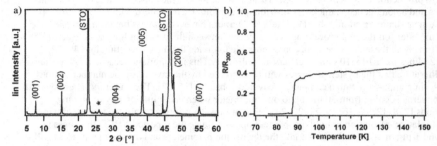

Figure 4. θ/2θ XRD spectrum (a) and resistivity measurements (b) of ink-jet printed YBCO patterns using the 60 μm Microfab nozzle on SrTiO$_3$ single crystal substrate after full thermal treatment.

Using the same 60 μm nozzle, the printing experiments were repeated on CeO$_2$-La$_2$Zr$_2$O$_7$-Ni-5at%W tape. Preliminary analysis has been performed on these samples. Optical profilometry revealed a line width of 170 ± 20 μm, and FIB-SEM cross-sectional analysis confirms the line thickness to vary between 220 to 250 nm, which is in good agreement with the predicted values using equation 2 and with the dimensions found for the same patterns printed on single crystal substrates.

CONCLUSIONS

Multi-filamentary YBCO patterns were successfully deposited on both SrTiO$_3$ and buffered Ni-5at%W tape using two set-ups of DOD piezo-electric ink-jet printers and a water-based YBCO ink. We are able to predict the pattern dimensions using knowledge of both the ink-jet printing parameters, jetting behavior, ink rheology and wettability on the selected substrates. The jettability of the ink was verified different nozzles by determining the inverse Ohnesorge number. After optimizing the inter-droplet distance and the number of successive coatings, tracks of 40 to 200 μm wide and 100 to 200 nm thick were obtained after full thermal processing. These tracks exhibit sharp edges and are very homogeneous in their overall profile, showing almost no coffee ring effect. After thermal processing, the patterns show strong c-axis oriented YBCO resulting in a T$_{c,onset}$ of 88.6 K and a ΔT$_c$ of 1.4 K when performing resistivity measurements on the 200 μm wide sample. Further optimization of the printing process will be performed to further increase the pattern thickness while keeping the width constant and avoiding random YBCO formation.

10

REFERENCES

1. Zhang, W.; Rupich, M. W.; Schoop, U.; Verebelyi, D. T.; Thieme, C. L. H.; Li, X.; Kodenkandath, T.; Huang, Y.; Siegal, E.; Buczek, D.; Carter, W.; Nguyen, N.; Schreiber, J.; Prasova, M.; Lynch, J.; Tucker, D.; Fleshler, S., Physica C **2007**, *463–465*, 505-509.
2. Iijima, Y.; Kakimoto, K.; Sutoh, Y.; Ajimura, S.; Saitoh, T., Physica C **2004**, *412–414, Part 2*, 801-806.
3. Shiohara, Y.; Kitoh, Y.; Izumi, T., Physica C **2006**, *445–448*, 496-503.
4. Van Driessche, I.; Feys, J.; Hopkins, S. C.; Lommens, P.; Granados, X.; Glowacki, B. A.; Ricart, S.; Holzapfel, B.; Vilardell, M.; Kirchner, A.; Baecker, M., SUST **2012**, *25* (6), 065017 (12p).
5. Selvamanickam, V.; Chen, Y.; Xiong, X.; Xie, Y. Y.; Martchevski, M.; Rar, A.; Qiao, Y.; Schmidt, R. M.; Knoll, A.; Lenseth, K. P.; Weber, C. S., IEEE Trans. Appl. Supercond. **2009**, *19* (3), 3225-3230.
6. Iijima, Y.; Hosaka, M.; Sadakata, N.; Saitoh, T.; Kohno, O.; Takeda, K., Appl Phys Lett **1997**, *71* (18), 2695-2697.
7. Cobb, C. B.; Barnes, P. N.; Haugan, T. J.; Tolliver, J.; Lee, E.; Sumption, M.; Collings, E.; Oberly, C. E., Physica C **2002**, *382* (1), 52-56.
8. Majoros, M.; Glowacki, B. A.; Campbell, A. M.; Levin, G. A.; Barnes, P. N.; Polak, M., IEEE Trans. Appl. Supercond. **2005**, *15* (2), 2819-2822.
9. Sumption, M. D.; Coleman, E. L.; Cobb, C. B.; Barnes, P. N.; Haugan, T. J.; Tolliver, J.; Oberly, C. E.; Collings, E. W., IEEE Trans. Appl. Supercond. **2003**, *13* (2), 3553-3556.
10. Carr, W. J.; Oberly, C. E., IEEE Trans. Appl. Supercond. **1999**, *9* (2), 1475-1478.
11. Oberly, C. E.; Razidlo, B.; Rodriguez, F., IEEE Trans. Appl. Supercond. **2005**, *15* (2), 1643-1646.
12. Amemiya, N.; Kasai, S.; Yoda, K.; Jiang, Z. N.; Levin, G. A.; Barnes, P. N.; Oberly, C. E., SUST **2004**, *17* (12), 1464-1471.
13. Oberly, C. E.; Long, L.; Rhoads, G. L.; Carr, W. J., Cryogenics **2001**, *41* (2), 117-124.
14. Tsukamoto, O.; Sekine, N.; Ciszek, M.; Ogawa, J., IEEE Trans. Appl. Supercond. **2005**, *15* (2), 2823-2826.
15. Glowacki, B. A.; Majoros, M., SUST **2000**, *13* (7), 971-973.
16. Sumption, M. D.; Barnes, P. N.; Collings, E. W., IEEE Trans. Appl. Supercond. **2005**, *15* (2), 2815-2818.
17. Goldacker, W.; Frank, A.; Heller, R.; Schlachter, S. I.; Ringsdorf, B.; Weiss, K. P.; Schmidt, C.; Schuller, S., IEEE Trans. Appl. Supercond. **2007**, *17* (2), 3398-3401.
18. Badcock, R. A.; Long, N. J.; Mulholland, M.; Hellmann, S.; Wright, A.; Hamilton, K. A., IEEE Trans. Appl. Supercond. **2009**, *19* (3), 3244-3247.
19. Suzuki, K.; Yoshizumi, M.; Izumi, T.; Shiohara, Y.; Iwakuma, M.; Ibi, A.; Miyata, S.; Yamada, Y., Physica C **2008**, *468* (15-20), 1579-1582.
20. Sumption, M. D.; Collings, E. W.; Barnes, P. N., SUST **2005**, *18* (1), 122-134.
21. Abraimov, D.; Gurevich, A.; Polyanskii, A.; Cai, X. Y.; Xu, A.; Pamidi, S.; Larbalestier, D.; Thieme, C. L. H., SUST **2008**, *21* (8), 082004 (4 p).
22. Duckworth, R. C.; Paranthaman, M. P.; Bhuiyan, M. S.; List, F. A.; Gouge, M. J., IEEE Trans. Appl. Supercond. **2007**, *17* (2), 3159-3162.

23. Minsoo, K.; Freyhardt, H. C.; Lee, T. R.; Jacobson, A. J.; Galstyan, E.; Usoskin, A.; Rutt, A., IEEE Trans. Appl. Supercond. **2013**, *23* (3), 6601304 (4 p).
24. Kopera, L.; Smatko, V.; Prusseit, W.; Polak, M.; Semerad, R.; Strbik, V.; Souc, J., Physica C **2008**, *468* (24), 2351-2355.
25. Glowacki, B. A.; Mouganie, T., Inst. Phys. Conf. Ser. **2003**, *No. 181*, 1884.
26. Feys, J.; Vermeir, P.; Lommens, P.; Hopkins, S. C.; Granados, X.; Glowacki, B. A.; Baecker, M.; Reich, E.; Ricard, S.; Holzapfel, B.; Van Der Voort, P.; Van Driessche, I., J. Mater. Chem. **2012**, *22*, 3717–3726.
27. Tekin, E.; Smith, P. J.; Schubert, U. S., Soft Matter **2008**, *4* (4), 703-713.
28. Windle, J.; Derby, B., J Mater Sci Lett **1999**, *18* (2), 87-90.
29. Derby, B., In *Annual Review Of Materials Research*, Annual Reviews: Palo Alto, 2010; Vol. 40, pp 395-414.
30. Arin, M.; Lommens, P.; Hopkins, S. C.; Pollefeyt, G.; Van der Eycken, J.; Ricart, S.; Granados, X.; Glowacki, B. A.; Van Driessche, I., Nanotechnology **2012**, *23* (16), 165603 (10p).
31. Mouganie, T.; Glowacki, B. A., J Mater Sci **2006**, *41* (24), 8257-8264.
32. Vermeir, P.; Feys, J.; Schaubroeck, J.; Verbeken, K.; Baecker, M.; Van Driessche, I., Mater. Chem. Phys. **2012**, *133* (2-3), 998-1002.
33. Van Driessche, I.; Penneman, G.; De Meyer, C.; Stambolova, I.; Bruneel, E.; Hoste, S.; Ttp, In *Euro Ceramics Vii, Pt 1-3*, 2002; Vol. 206-2, pp 479-482.
34. Penneman, G.; Van Driessche, I.; Bruneel, E.; Hoste, S., In *Euro Ceramics Viii, Pts 1-3*, Mandal, H. O. L., Ed. 2004; Vol. 264-268, pp 501-504.
35. Cloet, V.; Cordero-Cabrera, M. C.; Mouganie, T.; Glowacki, B. A.; Falter, M.; Holzapfel, B.; Engell, J.; Backer, M.; Van Driessche, I., Science and Engineering of Novel Superconductors **2006**, 153-158.
36. Vermeir, P.; Deruyck, F.; Feys, J.; Lommens, P.; Schaubroeck, J.; Van Driessche, I., J. Sol-Gel Sci. Techn. **2012**, *62* (3), 378-388.
37. Lommens, P.; Feys, J.; Vrielinck, H.; De Buysser, K.; Herman, G.; Callens, F.; Van Driessche, I., Dalton Trans. **2012**, *41* (12), 3574-3582.
38. Fromm, J. E., IBM J Res Dev **1984**, *28* (3), 322-333.

Mater. Res. Soc. Symp. Proc. Vol. 1547 © 2013 Materials Research Society
DOI: 10.1557/opl.2013.856

'in-situ' preparation of metal oxide thin films by inkjet printing acetates solutions

Mei Fang,[1,2] Wolfgang Voit,[2] Yan Wu,[3] Lyubov Belova[2] and K.V. Rao[2]

[1] Department of Physics, Fudan University, Shanghai 200433, China
[2] Department of Materials Science and Engineering, KTH-Royal Institute of Technology, Stockholm, SE10044, Sweden
[3] Faculty of Materials Science and Chemical Engineering, China University of Geosciences, Wuhan, 430074, China

ABSTRACT

Direct printing of functional oxide thin films could provide a new route to low-cost, efficient and scalable fabrications of electronic devices. One challenge that remains open is to design the inks with long term stability for effective deposition of specific oxide materials of industrial importance. In this paper, we introduce a reliable method of producing stable inks for 'in-situ' deposition of oxide thin films by inkjet printing. The inks were prepared from metal-acetates solutions and printed on a variety of substrates. The acetate precursors were decomposed into oxide films during the subsequent calcination process to achieve the 'in-situ' deposition of the desired oxide films directly on the substrate. By this procedure we have obtained room temperature contamination free ferromagnetic spintronic materials like Fe doped MgO and ZnO films from their acetate(s) solutions. We find that the origin of magnetism in ZnO, MgO and their Fe-doped films to be intrinsic. For a 28 nm thick film of Fe-doped ZnO we observe an enhanced magnetic moment of 16.0 emu/cm^3 while it is 5.5 emu/cm^3 for the doped MgO film of single pass printed. The origin of magnetism is attributed to cat-ion vacancies. We have also fabricated highly transparent indium tin oxide films with a transparency >95% both in the visible and IR range which is rather unique compared to films grown by any other technique. The films have a nano-porous structure, an added bonus from inkjetting that makes such films advantageous for a broad range of applications.

INTRODUCTION

As an efficient, inexpensive and scalable technique, inkjet printing offers an ideal answer to the emerging trends and demands of depositing small volume (in picoliter range) droplets of precursor liquid inks into functional thin films and device components with a high degree of pixel precision. Compared to other techniques of film deposition, this method is fast, simple, precise, material-saving and suitable for any type of substrates, a promising advantage for new flexible and/or stretchable electronics. Thus, of late, inkjet technology have been used for fabrications of different materials for various applications, including organic transistor circuits,[1-5] polymer displays,[6] metallic nanoparticles for flexible electronics,[7-9] organic solar cells,[10-12] patterning of biomolecules, cells and tissues for biomedicine applications[13, 14], and many more.[15, 16] One of the remaining challenges of developing inkjet materials deposition for these applications is the fabrication of inks suitable for printing inorganic materials. Some specific physicochemical properties of the inks for inkjet printing are, for instance, the viscosity (η) should be in range of 1~25 mPa·s and the surface tension (γ) should range in 20~50 mN·m^{-1} for drop on demand (DOD) inkjet printer.[17] Besides, the ink should be stable without sediments and should not chemically react with the printing system.

For some metallic particles with nano-size such as Ag and Au, [7, 8] colloids have been prepared by suspending the particles in organic solvent utilizing Brownian motion and/or surface modification, to render them suitable as inkjet inks. The printing process transports the particles from the ink to the substrate, and post-annealing is required to evaporate the solvent and sinter the nanoparticles together. This method could not be applied for particles with larger size, since their gravity is too large to be overcome by the interactions among the particles and therefore could settle in a short time, i.e., the suspension is not stable. For metal oxides, the high metallurgical bonding energy requires very high post-annealing temperature, which limits the applications of inkjet-printing in many electronic devices. Recently, nitrates and hydroxides were used to prepare ink precursors for printing,[18, 19] which could 'in-situ' synthesize oxide thin films, i.e., the oxides are formed at a given site on the substrate.

Since acetates have a low decomposition temperature range and they are soluble in water and many organic solvents, we used metal-acetates to prepare inks for printing different oxide thin films. This idea originates from the sol-gel process, which is well reported. Fan and Boettcher used sol-gel solution consisting of acetic acid, hydroxchloric acid and ethanol, and prepared membrane of metal oxide through catalytic reaction.[20-22] Liu developed colloidal nanoparticle inks for inkjet printing metal oxide.[23] These studies prepared metal oxides 'in-situ' on substrate via catalytic reaction of the solution, followed by the calcination of the organics. In our work, we used metal-acetates as precursors. The introduced elements in the ink solution are carbon, hydrogen, oxides. They can be easily burned out to prevent any contamination of the products. Metal oxides are prepared from the decomposition of metal-acetates 'in-situ' on substrates during the calcination process, which enables the tuning of film structure during the thermal treatment. Since the inks are single phase solutions, they could be stable for months without sediments. By selecting the solvents, the physicochemical properties of the inks could be adjusted in the range suitable for inkjet printing. We have prepared acetate precursor inks for printing (1) ZnO and MgO thin films, (2) Fe-doped ZnO and MgO thin films, which show robust room temperature ferromagnetism; and also (3) Indium tin oxide (ITO) thin films which have transparency > 95% both in visible and IR range. The films obtained are uniform, and the printing process is reliable and repeatable.

EXPERIMENT

Ink preparation: The inks were prepared by dissolving metal-acetates in organic solvents. (1) Inks for printing ZnO and Fe-doped ZnO films: Zinc acetate dehydrate $(Zn(OAc)_2 \cdot 2H_2O$ Alfa Aesar) was dissolved into 2-isopropoxyethanol (IPE, $(CH_3)_2CHO(CH_2)_2OH$ from Sigma-Aldrich with $\eta=\sim2.4$ mPa·s and $\gamma=\sim28$ mN·m^{-1}), which was used as ink precursors for printing pure ZnO films. For Fe-doped ZnO films, the ink was prepared by dissolving $Zn(OAc)_2 \cdot 2H_2O$ and iron (II) acetate anhydrous $(Fe(OAc)_2$ from Alfa Aesar) in IPE with designed ratio of Fe^{2+} to Zn^{2+}. (2) Similar as the ink preparations for printing ZnO and Fe-doped ZnO thin films, inks of MgO and Fe-dope MgO thin films were prepared by dissolving magnesium acetate tetrahydrate $(Mg(OAc)_2 \cdot 4H_2O$ from Alfa Aesar) and $Fe(OAc)_2$ in methoxyethanol (MOE, $CH_3O(CH_2)_2OH$, ACS 99.3% from Alfa Aesar with $\eta=\sim1.7$ mPa·s and $\gamma=\sim33$ mN·m^{-1}). Drops of acetic acid were used to improve the solubility of acetate salts in MOE. (3) Inks for printing ITO films were prepared by dissolving indium (III) acetate

(In(OAc)$_3$, 99.99%, Alfa Aesar) and tin (IV) acetate (Sn(OAc)$_4$ Sigma-Aldrich) in acetylaceton (ATA, CH$_3$COCH$_2$COCH$_3$, ≥99% from Sigma-Aldrich with η=~0.8 mPa·s and γ=~31 mN·m^{-1}), with cation ratio of [Sn^{4+}]:[In^{3+}]=1:9. The solution was heated on a hotplate kept at 120 °C and 0.01 mol hydrogen peroxide (30 wt.% in H$_2$O, Merck) was added dropwise in 6 aliquots at 30 minutes intervals to improve the solubility and chelate the precursor salts. The concentration of all the inks can be adjusted to achieve ideal thickness of single pass printed oxide films. The prepared inks were characterized by thermogravimetric analysis (TG, Perkin-Elmer TGS-2) with heating rate of 20 °C/min to detect the thermal processing conditions after printing.

Substrate preparation: The substrates we used were glass and silicon, which were cleaned well for 10 minutes by ultrasonic agitation in acetone and another 10 minutes in ethanol, and then soaked in isopropanol. They were dried by nitrogen gas flow and preheated before printing. The substrate temperature was 60°C for printing ZnO and Fe-doped ZnO films, and 80 °C for printing MgO, Fe-doped MgO, and ITO films, with optimized evaporation rate of liquid in the inks for deposition of uniform films without the formation of the so called 'coffee-rings'.

Inkjet printing: The inks were printed on pre-heated substrates by a piezoelectric shear mode drop on demand inkjet printer, using XJ 126/50 printhead from Xaar. For each printed layer, the deposited ink was dried under the substrate temperature, and the solid precursors were then heated on a 150°C hotplate for 10 minutes to condense the film. For multi-layer printed films, a new pass of printing was deposited on top of the condensed precursors, followed by the same route of drying and condensing. The precursor films were annealed in a pre-heated furnace to transform the metal acetates into oxides. The ZnO and Fe-doped ZnO films were prepared from thermal annealing at 450 °C for 1 hour; the MgO and Fe-doped MgO films were prepared from annealing at 450°C for 2 hours and then 600 °C for 2 hours; the precursors of ITO films were annealed at 500 °C for 2 hours.

Characterizations: The morphology of the surface and the cross-section of the inkjet printed films were studied in a dual beam combined focused ion beam / scanning electron microscope (FIB/SEM, FEI Nova 600 Nanolab). Elemental analysis of printed films were determined by energy dispersed X-ray spectroscopy (EDXS) attached to the FIB/SEM system. The structure of the printed films was investigated in a copper anode X-ray diffractometer (XRD, Siemens D5000). A superconducting quantum interference device (SQUID, Quantum Design MPMS2) was used to determine the magnetic properties of ZnO, MgO and their Fe-doped films. For ITO films, the electronic transport properties were characterized by I-V curves determined from a home-designed set up using a standard four-probe method. The optical properties were studied by means of a Woollam M-2000 ellipsometer.

DISCUSSION

Figure 1 shows the visual appearance of the inks for inkjet printing, which were prepared by dissolving acetates in organic solvents with suitable viscosity and surface tension. The inks are single phase and transparent with colors corresponding to the different types of ions. These inks are stable overtime without sediments, and compatible with the printing system without any undesirable chemical reactions.

Figure 1. *Inks prepared from metal acetates for printing different oxide films: ZnO, Fe-doped ZnO, MgO, Fe-doped MgO, and ITO.*

The prepared inks were characterized by thermo-gravimetric analysis to determine the conditions for annealing after printing. Figure 2 shows the weigh changes of the inks with the temperature. The inks as-prepared show a dramatic weight loss at ~150°C, ~150°C, and ~135°C for (a) Fe-doped ZnO, (b) Fe-doped MgO, and (c) ITO, respectively. These temperatures correspond to the evaporations of solvents in the ink, i.e., IPE, MOE, and ATA for (a), (b) and (c), respectively. After being dried on a hotplate at 150°C for 10 minutes, the liquid in the inks evaporated and the solid precursors were left for thermo-gravimetric analysis, shown as the curves of 'ink-after being dried' in the figure. The decomposition temperature range of the solid precursors is 220~440 °C for Fe-doped ZnO, 320-450 °C for Fe-doped MgO, and 165~445 °C for ITO, respectively. The weight loss observed at 240 °C for the dried ink of Fe-doped MgO could be due the loss of crystallization water in the precursors. The weight percent of products from precursors is found to be consistent with the weight changes expected from metal-acetates to their corresponded oxides.[24]

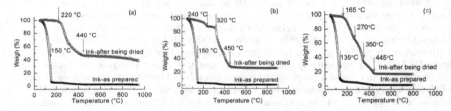

Figure 2. *Thermo-gravimetric analysis of the inks prepared for printing (a) Fe-doped ZnO, (b) Fe-doped MgO and (c) ITO thin films.*

Figure 3 shows the schematic process of inkjet printing oxide thin films from 'ex-situ' and 'in-situ' processes. For the 'ex-situ' process, the oxide particles were first synthesized and dispersed in solvent to prepare colloid or suspension for inkjet printing. The printing process transports the oxides particles to a substrate and hence the properties of the films depend a lot on the morphology of the particles in addition to the printing process. The films are typically fragile with weak correlations among the particles, and their density and continuity are limited, as seen in a typical SEM image shown in the insert in Fig. 3(a).[25] To achieve atomic bonding, high temperature annealing is needed to sinter the oxide particles on the substrate together and form films. For the 'in-situ' process, the precursors were printed on the substrate and transformed into oxide during annealing. The oxides were prepared in position on the substrate, which enables the structure tuning and ensures the quality of the oxide films. Since acetates have a low decomposition temperature (typically lower than 450 °C, see examples shown in Fig.2) and the introduced elements, i.e. carbon and hydrogen, can be burned without impurities, high quality oxide films could be formed at a low temperature. The typical SEM image of 'in-situ' films is shown in the insert figure in Fig. 3(b). Besides, by controlling the annealing conditions, the structure of the films could be tuned. [26]

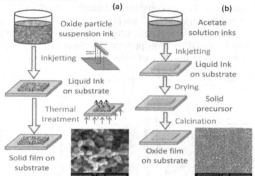

Figure 3. *Schematic diagrams of (a) 'ex-situ' and (b) 'in-situ' inkjet printing processes. The insert SEM images are typical morphology of films from 'ex-situ' and 'in-situ' processes, respectively.*

Since the oxide films are transferred from acetate precursors during the annealing process after 'in-situ' inkjet printing, it is possible to dope secondary element into the matrix oxide lattice without forming impurity oxide phases. Figure 4 shows elemental and structural X-ray analysis of 10 at.% Fe-doped ZnO films which were prepared from a ink with [Fe]:[Zn]= 1:9. Fe element is clearly determined in EDXS, while in XRD data the diffraction peaks are matched well with pure ZnO without any impurity oxide phases. This indicates that the Fe atoms are doped into ZnO lattice without forming iron oxides. This 10 at.% Fe-doping in ZnO system is much higher than the doping limit of 7% reported from literature using mechanical alloying. [27]

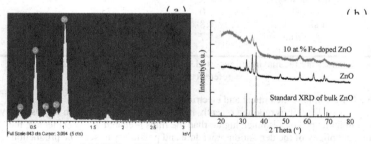

Figure 4. *(a) EDXS and (b) XRD of 10 at.% Fe-doped ZnO thin films prepared from 'in-situ' inkjet printing process.*

By 'in-situ' inkjet printing, we have prepared secondary element doped ZnO and MgO films without impurity oxide phases, and can be used in spintronic and electronic applications. Figure 5 shows a typical magnetic hysteresis loop of the films determined at room temperature, which shows robust room temperature ferromagnetism with a coercivity of ~75 Oe. It was found that for comparable film thickness (single pass printed), Fe-doping could enhance the magnetic moment by 16.0 emu/cm^3 and 5.5 emu/cm^3 for ZnO and MgO systems, respectively. More details will be published elsewhere. Table I compares the magnetism of 'in-situ' ink-jet printed ZnO, MgO, Fe-doped ZnO and Fe-doped MgO films prepared by different deposition methods.[28-30] The results indicate that the magnetic properties of 'in-situ' prepared oxide films can be even better than those observed for films prepared by physical vapor deposition technique for example. It could be related to the structure of films from different methods.

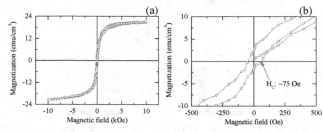

Figure 5. *Typical magnetic hysteresis loop measured at room temperature for ink-jetted films: (a) the whole loop for a 28 nm thick Fe-doped ZnO thin film and (b) the enlarged view of the loop around low fields for coercivity determination.*

Table I. Magnetic properties of 'in-situ' inkjet printed films compared with the reported values of the films prepared by different deposition techniques.

Materials	Deposition methods	Ms (emu/ cm^3)	Film thickness (nm)	References
ZnO	'in-situ' inkjet	5.0	30	
ZnO	sputtering	3.5	48	28
ZnO	PLD	2.0	375	29
Fe$_{0.1}$Zn$_{0.9}$O	'in-situ' inkjet	24.9	45	
Fe$_{0.1}$Zn$_{0.9}$O	'in-situ' inkjet	21.0	28	
Fe$_{0.01}$Zn$_{0.99}$O	PLD	2.0	375	29
MgO	'in-situ' inkjet	0.8	90	
MgO	sputtering	5.7	170	30
Fe$_{0.1}$Mg$_{0.9}$O	'in-situ' inkjet	6.3	30	

Note: the Ms values are the maximum reported saturation magnetization in the indicated reference.

In table II we compare the optical and electrical properties of our 'in-situ' printed ITO films with the 'ex-situ' ITO films reported recently in literature.[31-33] Even though the resistivity of ITO films from inkjet printing is much higher than the films prepared by PVD method, the material-saving, simplicity of the deposition procedure and possibility to scalable production make inkjet technology still attractive for printing ITO films. By 'in-situ' printing, we could prepare very thin (~40 nm) ITO films with resistivity comparable to that of thick films (~580 nm) by an 'ex-situ' process. Also, the transparency of our ITO films is higher than 95% in the visible range, which is rather unique compared to the properties of films obtained from other techniques.

Table II Electrical and optical properties of ITO films prepared from 'ex-situ' and 'in-situ' inkjet printing.

Methods	Resistivity (Ωcm)	Transparency	Annealing conditions	Film thickness (nm)	References
'in-situ'	2.9×10^{-2}	>95%	500 °C	40	
'ex-situ'	1.5×10^{-2}	85.9%	450 °C	800	[31]
	3.0×10^{-2}	87%	400 °C	580	[32]
	1.5×10^{-2}	86.7%	450 °C	750	[33]

Note: The properties listed here are from the optimized samples present in the references.

CONCLUSIONS

We have prepared uniform and dense ZnO, MgO, Fe-doped ZnO, Fe-doped MgO and ITO films by 'in-situ' inkjet printing. The inks were prepared from acetate solutions, which give films that are of single phase and long term stability. Because of the 'in-situ' process, we achieved secondary elemental doping up to 10 at.% in the oxide films without forming impurity oxide phases. The properties of these 'in-situ' prepared films are comparable to those of films obtained by PVD method. For instance, a 45 nm thick Fe-doped ZnO film has magnetic moment of 24.9 emu/cm^3 (~0.83 μ_B/Fe atom), which is much higher than the reported values from literature. Another example for the advantages of our acetate solutions approach is illustrated by fabricating ITO films exhibiting high transparency (>95%) which is rather unique compared to films grown by other techniques. The 'in-situ' inkjet materials deposition using acetate precursors is a quite suitable method for producing functional oxide films for various applications in spintronics, electronics and optics.

ACKNOWLEDGMENTS

Mei Fang was supported by China Scholarship Council for her Ph.D. study in Sweden. This project at RIT was funded by the Swedish Agency VINNOVA.

REFERENCES

1. H. Sirringhaus, T. Kawase, R. H. Friend, T. Shimoda, M. Inbasekaran, W. Wu, and E. P. Woo, Science **290**, 2123 (2000).
2. T. Kawase, T. Shimoda, C. Newsome, H. Sirringhaus, and R. H. Friend, Thin Solid Films **438-439**, 279 (2003).
3. N. Stutzmann, R. H. Friend, and H. Sirringhaus, Science **299**, 1881 (2003).
4. T. Sekitani, Y. Noguchi, U. Zschieschang, H. Klauk, and T. Someya, Proceedings of the National Academy of Sciences of the United States of America **105**, 4976 (2008).
5. H. Yan, Z. Chen, Y. Zheng, C. Newman, J. R. Quinn, F. Dötz, M. Kastler, and A. Facchetti, Nature **457**, 679 (2009).
6. T. Shimoda, K. Morii, S. Seki, and H. Kiguchi, MRS Bulletin **28**, 821 (2003).

7. D. Huang, F. Liao, S. Molesa, D. Redinger, and V. Subramanian, Journal of the Electrochemical Society **150**, G412 (2003).
8. H. H. Lee, K. S. Chou, and K. C. Huang, Nanotechnology **16**, 2436 (2005).
9. D. Tobjörk and R. Österbacka, Advanced Materials **23**, 1935 (2011).
10. E. A. Roth, T. Xu, M. Das, C. Gregory, J. J. Hickman, and T. Boland, Biomaterials **25**, 3707 (2004).
11. C. N. Hoth, P. Schilinsky, S. A. Choulis, and C. J. Brabec, Nano Letters **8**, 2806 (2008).
12. S. H. Eom, S. Senthilarasu, P. Uthirakumar, S. C. Yoon, J. Lim, C. Lee, H. S. Lim, J. Lee, and S. H. Lee, Organic Electronics: physics, materials, applications **10**, 536 (2009).
13. T. Xu, J. Jin, C. Gregory, J. J. Hickman, and T. Boland, Biomaterials **26**, 93 (2005).
14. I. Barbulovic-Nad, M. Lucente, Y. Sun, M. Zhang, A. R. Wheeler, and M. Bussmann, Critical Reviews in Biotechnology **26**, 237 (2006).
15. P. Calvert, Chemistry of Materials **13**, 3299 (2001).
16. M. Singh, H. M. Haverinen, P. Dhagat, and G. E. Jabbour, Advanced Materials **22**, 673 (2010).
17. S. Magdassi, *The chemistry of inkjet inks* (World Science Publising Co. Pte. Ltd., Singapore, 2010).
18. G. H. Kim, H. S. Kim, H. S. Shin, B. D. Ahn, K. H. Kim, and H. J. Kim, Thin Solid Films **517**, 4007 (2009).
19. C.-C. Huang, P.-C. Su, and Y.-C. Liao, Thin Solid Films.
20. J. Fan, S. W. Boettcher, and G. D. Stucky, Chemistry of Materials **18**, 6391 (2006).
21. S. W. Boettcher, J. Fan, C. K. Tsung, Q. Shi, and G. D. Stucky, Accounts of Chemical Research **40**, 784 (2007).
22. J. Fan, Y. Dai, Y. Li, N. Zheng, J. Guo, X. Yan, and G. D. Stucky, Journal of the American Chemical Society **131**, 15568 (2009).
23. X. Liu, et al., Nano Letters **12**, 5733 (2012).
24. M. Fang, W. Voit, A. Kyndiah, Y. Wu, L. Belova, and K. Rao, MRS Proceedings **1394** (2012).
25. E. Girgis, M. Fang, E. Hassan, N. Kathab, and K. V. Rao, Journal of Materials Research **28**, 502 (2012).
26. Y. Wu, Y. Zhan, M. Fahlman, M. Fang, K. Rao, and L. Belova, MRS Proceedings **1292** (2011).
27. Y. Lin, D. Jiang, F. Lin, W. Shi, and X. Ma, Journal of Alloys and Compounds **436**, 30 (2007).
28. M. Kapilashrami, J. Xu, V. Ström, K. V. Rao, and L. Belova, Applied Physics Letters **95** (2009).
29. N. H. Hong, E. Chikoidze, and Y. Dumont, Physica B: Condensed Matter **404**, 3978 (2009).
30. C. M. Araujo, et al., Applied Physics Letters **96** (2010).
31. H.-K. Kim, I.-K. You, J. B. Koo, and S.-H. Kim, Surface and Coatings Technology **211**, 33 (2012).
32. M.-s. Hwang, B.-y. Jeong, J. Moon, S.-K. Chun, and J. Kim, Materials Science and Engineering: B **176**, 1128 (2011).
33. J.-A. Jeong and H.-K. Kim, Current Applied Physics **10**, e105 (2010).

Mater. Res. Soc. Symp. Proc. Vol. 1547 © 2013 Materials Research Society
DOI: 10.1557/opl.2013.507

Growth and physical properties of vanadium oxide thin films with controllable phases

Yanda Ji[1,] Yin Zhang[1], Min Gao[1], Zhen Yuan[2], Changqing Jin[2], Yuan Lin[1]*

1. State Key Laboratory of Electronic Thin films and Integrated Devices, University of Electronic Science and Technology of China, Chengdu, Sichuan 610054, P. R. China
2. The Institute of Physics, Chinese Academy of Sciences, P.O. Box 603, Beijing 100190, P. R. China

ABSTRACT

Vanadium oxides thin films with variable oxidation states have attracted great attention due to their unique electrical, optical properties and many important applications in microelectronics, infrared optical devices, and energy harvest systems. However, to fabricate vanadium oxide thin films with controllable phases and desired transport properties is still a challenge by using a chemical solution deposition (CSD) technique. In this paper, we report that vanadium oxide thin films with well controlled phases such as rhombohedral V_2O_3 and monoclinic VO_2 could be synthesized on Al_2O_3 (0001) substrates using a CSD technique ---- polymer assisted deposition (PAD). Both V_2O_3 and VO_2 thin films can be well controlled with good epitaxial quality by optimizing the fabrication parameters. The electrical resistivity changes 3~4 orders of magnitude at metal insulator transition for both epitaxial V_2O_3 and VO_2 thin films. The correlation between the physical properties and the microstructures of the films will be discussed.

INTRODUCTION

The vanadium element exhibits different chemical valences in the family of vanadium oxides including V_2O_3, VO_2 and V_2O_5 etc. V_2O_3 is always considered to be a typical Mott insulator and undergoes Mott metal insulator transition (MIT) above the liquid nitrogen temperature [1].This unique property makes V_2O_3 to be a great candidate to investigate the theory of Mott transition mechanism. However, there is few report on the device applications based on V_2O_3 due to its low transition temperature at around 155 K. Fortunately, VO_2 exhibits the MIT property at just above room temperature. Under this temperature, many feasible devices applications in microelectronics, infrared optical devices, and energy harvest systems can been developed [2-4]. In fact, the structure of VO_2 changes from a monoclinic, called M1 phase, to a rutile structure with the intermediate monoclinic M2 phase [5]. As a result, the metal insulator transition mechanism is under a controversy of Peierls or Mott mechanisms [6].

Because of difficulties in controlling the multiple chemical valences of vanadium, it is still a challenge to deposit vanadium oxide thin films with controllable phases and desired transport properties, especially by using a low-cost and manufacturable way. Compared to other thin film deposition techniques, chemical solution deposition (CSD) techniques show advantages of low cost, easy setup and ability for large area coating. Polymer assisted deposition (PAD) is an effective chemical solution deposition technique for the fabrication of functional oxide thin films [7]. In this paper, we will report that both epitaxial V_2O_3 and VO_2 thin films can be grown on Al_2O_3 (0001) substrates by the polymer assisted deposition technique, through the precise control of the thermal treatment process. MIT properties of the films have been investigated and suggested to be related with the microstructures in the films.

EXPERIMENTAL DETAILS

The precursor solution used in our experiments was prepared using polyethyleneimine (PEI), a type of water-soluble polymer, which has a great deal of amino groups in the polymer molecules. These groups are capable of coordinating to metal ions to form amino chelates. For some typical metal ions (such as Cu^{2+}, Cd^{2+} and Zn^{2+}), PEI displays an excellent chelating ability together with fine hydrophilicity of the precursor [8]. However, not all metal ions are sensitive to amino chelating. For some types of metal ions, ethylene diamine tetraacetic acid (EDTA) could be introduced into the polymer system to realize a good binding between the metal ions and the polymer chains. EDTA is commonly known as carboxylic acid complex agent, and widely chelates with the majority of metal ions. Carboxylate radical in EDTA takes a reaction with ammonia radical in PEI in the form of acid and alkali neutralization reaction. In this way, almost all of the metal and metal acid radical ions (alkaline-earth metal ions excluded) can be bound to the polymer to form the polymeric precursor solutions. In this experiment, 2.0 g of PEI (Sigma-Aldrich, average Mn≈60000, Mw≈750000) was dissolved in 40 mL of distilled water, then 3.5 g EDTA (AR, Aladdin Chemistry Co. Ltd) was added into the previous solution to form PEI-EDTA water solution. As expected, neutralization reaction happened between the acid and alkali so that the EDTA can be considered to anionic graft to the polymer chain of PEI. For preparing the V polymer precursor solution, 1.4 g NH_4VO_3 (AR, Aladdin Chemistry Co. Ltd) was dissolved in the PEI-EDTA solution, and the VO_3^- ions were bound to the polymer chains. After that, the solution was purified and concentrated in an Amicon filtration unit (Amicon 8050) with the filter membrane. The final concentration of vanadium in the as-prepared solution, measured by inductively coupled plasma optical emission spectrometer (ICP-OES, Varian 700, Agilent Technologies), was 0.37 mmol/mL. Then, the solution was spin-coated on the Al_2O_3 (0001) substrates with a spin rate of 5000 rpm for 30 s. The thickness of the prepared thin film is controlled by the times of multiple spin coating.

The as-prepared precursor films were then put into a furnace for a heat treatment to depolymerize the polymer and crystallize the films. The conditions such as temperature and ambient of heat treatment needed to design based on the thermal dynamic phase diagram of vanadium oxides. Details will be discussed in next session.

The crystalline structures of the as-grown thin films were characterized using a Bede D1 X-ray diffract meter with CuKα radiation (λ=1.54060 Å)). Further studies on microstructures were carried out by Raman spectra recorded at various temperatures using a Renishaw Micro-Raman Spectroscopy System. The MIT properties of the thin films were investigated by 4 points resistance test from Agilent B2901A and the temperature was controlled by Lakeshore 325.

RESULTS AND DISCUSSION

The structure of V_2O_3 is rhombohedral and the a-axis lattice parameter is 4.901 Å, which is very close to that of Al_2O_3 substrate (a=4.760 Å, JCPDS 89-3072). Thus, V_2O_3 may epitaxially grow on Al_2O_3 (0001) substrate with its (0001) axis normal to the surface of the substrate. On the other hand, the crystalline structure of VO_2 is monoclinic. The VO_2 thin films could be epitaxially grown on the Al_2O_3 (0001) substrates with the out-plane orientation is [020] or [002], which depends on the tilting direction of the VO_2 dimers in the thin films.

To control the oxidation states of the vanadium oxides, the annealing atmosphere should be carefully controlled. The vanadium in the V_2O_3 has a relatively low valence (+3), which requires a low partial oxygen pressure during its growth. The VO_2 has an intermediate oxidation state whereas the V_2O_5 has the highest oxidation state. Thus, a schematic thermal dynamic diagram can be drafted as figure 1, which would help us design the heat treatment process for vanadium oxides with different oxidation states.

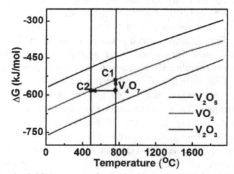

Figure 1. The standard Gibbs energy formation for V_2O_3, VO_2 and V_2O_5, respectively.

According to the thermochemical theory, the balanced chemical reaction thermodynamic equation can be described as:

$$\Delta_r G_m^\theta = -gRT ln\frac{\gamma p_{O_2}}{p^\theta}$$

where $\Delta_r G_m^\theta$ is the standard Gibbs energy, g is equilibrium coefficient, R is a universal constant, T is the temperature, p_{O_2} is the partial pressure of oxygen, p^θ is standard atmospheric pressure and γ is the fugacity coefficient which can indicate the pressure effects. It should be noted that the partial pressure of oxygen is determined once the atmosphere is chosen, according to the Dalton law of additive pressure. And the fugacity coefficient γ for oxygen decreases as the pressure of reducing atmosphere increases. That means the effective partial pressure of oxygen decreases and a lower chemical valence vanadium oxide can be produced. Thus, to get the low oxidation state in V_2O_3, we have tried to synthesis the films in a reduced ambient under a high pressure. Technically, the preparation of epitaxial V_2O_3 thin films on Al_2O_3 substrates by PAD technique is as following. After spin coating, the precursor thin films were annealed in a high pressure tube furnace (MTI OTF 1200x) in a reduced atmosphere (2% H_2 and 98% N_2) under the pressure of 0.3 MPa. The samples was annealing at 750℃ for 4 hours. X-ray diffraction (XRD) θ-2θ scans and φ scans were performed. The results are shown in figure 2, the V_2O_3 thin film was verified to be epitaxially grown on the Al_2O_3（0001）substrate with the interface relationship of V_2O_3 (0001)//Al_2O_3 (0001) and V_2O_3 [11-20]//Al_2O_3 [11-20]. To achieve the growth of single phase VO_2, lower reducing environment is required when compared to the case of V_2O_3. Thus, we reduce the pressure of the forming gas (2% H_2 and 98% N_2) to 0.1 MPa (~ 1 atm) to reduce the fugacity coefficient γ of this kind of reducing atmosphere. However, it is found that we could only get V_4O_7 rather than VO_2 at 750℃. Based on figure 1, to further increase the oxidation state of VO_x, we can either induce a small amount of oxygen or reduce the annealing temperature. Experimental data showed that both ways are feasible. A heat treatment of the films in a weak oxidizing environment (Ar 99% and O_2 1%) at 750 °C for 60 min, right after depolymerizing the precursor films in a reduced environment (N_2 98 % and H_2 2%) could ensure the growth of epitaxial single phase VO_2 films, as demonstrated in our previous report [9]. This condition can be indicated as point C1 in figure 1. On the other hand, another optimized processing condition has been found by reducing the annealing temperature, as indicated by point C2 in figure 1. After spin coating, the precursor films were heated in a tube furnace (MTI GSL 1500X) to 470℃ to depolymerize the polymer, then stayed there for 60 min to crystallize the thin films before cooling down to room temperature. The whole process was performed in the flowing forming gas (2% H_2 and 98% N_2) with the flowing rate of 150 mL/min. Such a strategy could simplify the synthesis steps and reduce the annealing temperature for the growth of VO_2 by PAD technique.

The microstructures of the as-prepared VO_2 thin films made by using the strategy of reducing annealing temperature, i.e. condition C2 in figure 1, were also characterized by the XRD technique. The typical results are shown in figure 3. As reported in the literatures, there are two opinions in determining the epitaxial behavior of VO_2 on Al_2O_3 (0001) substrate, i.e. the normal axis of the film is along VO_2 [001] or VO_2 [010]. As can be seen from the fact that the (100) and (010) orientation of VO_2 are equivalent in the rutile phase, the vanadium atoms dimerize along c axis and tilt across the phase transition. Based on our current XRD or TEM data, it is still challenge to tell which mode is dominant in our case, since the featured diffraction peaks in both cases are too close to each other in either the normal or tilted scans. So all the discussion below is based on the hypothesis that VO_2 (001)//Al_2O_3 (0001). In the pattern from the θ-2θ scan (figure 2(c)), only peaks from VO_2 (00l) can be seen besides the peaks from the substrate. The peak from VO_2 (001) centers at 39.74°, which is larger than that of the bulk VO_2, indicating a shorter lattice parameter in the Z-axis compared to VO₂ bulk and residual stress should exist. XRD φ scans were performed to study the epitaxial behavior of the as-prepared VO₂ films. Figure 2(d) shows the φ scan patterns from VO₂ {011} and Al₂O₃ {10-12}. As we know, VO₂ has a monoclinic structure with 1-fold symmetry while the structure of Al₂O₃ can be considered as hexagonal with six-fold symmetry. In figure 2(d), six peaks from VO₂ {011} can be observed, indicating that there are six possible in-plane variants which are rotated by 60° with each other around the axis normal to the surface of the film. The average value of FWHM of the peaks from VO₂ {011} is about 2.15°, demonstrating that the film has good epitaxial quality. The interface relationships between the thin film and substrate can be regarded as VO₂ [010]//Al₂O₃ [11-20] and VO₂ (001)//Al₂O₃ (0001).

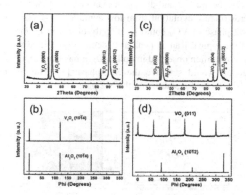

Figure 2. (a) XRD θ-2θ scans for as-prepared V₂O₃ thin film on Al₂O₃ (0001) and (b) φ scans for as-prepared V₂O₃ thin film on Al₂O₃ (0001); (c) XRD θ-2θ scans for as-prepared VO₂ thin film on Al₂O₃ (0001) and (d) φ scans for as-prepared VO₂ thin film on Al₂O₃ (0001).

Both of the MIT properties for as-prepared V₂O₃ and VO₂ thin films were studied by 4 points resistance test at various temperatures by using Agilent B2901A while the temperature was controlled by Lakeshore 325. The results are shown in figure 3. The resistance of V₂O₃ thin films exhibit four orders of magnitude changing from 1.86 MΩ at 80 K to 43.58 Ω at 200 K. Similarly the VO₂ thin film also shows a sharp MIT transition at 341K with three orders of magnitudes changing in the resistance from 0.24 MΩ at 300 K to 18.89 Ω at 375 K. Both resistance-temperature curves of the V₂O₃ and VO₂ films show loop curves, indicating the nature of latent heat for the first order phase transition of the as-prepared vanadium oxide thin films.

To further study the microstructure evolution of as-prepared VO₂ thin films across the MIT process, Raman spectrum was carried out at various temperatures. As shown in figure 4(a), at room temperature, the as-grown VO₂ thin film exhibits the same curve as the M1 phase of bulk single crystalline VO₂, indicating the thin film has a good crystalline structure. As the temperature rising up (Figure 4(b)), the mapping peaks of Raman spectrum originated from M1

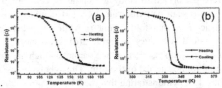

Figure 3. (a) MIT properties of as prepared V_2O_3 thin films; (b) MIT properties of as prepared VO_2 thin films.

phase of VO_2 thin films became weaker and disappeared at last. This phenomenon is considered to correspond to the microstructure evolution from low symmetry M1 phase to high symmetry rutile phase. Specially, M2 phase of VO_2 could be directly observed during the temperature of 334K, with the feature of the typical Raman peak at 618cm^{-1} blue shifted to 642cm^{-1}. Although M2 phase is metastable and not often observed in bulk VO_2, it has been reported that the uniaxial stress along $[110]_R$ direction would stabilize the M2 phase [5]. The residual stress in the as-grown thin film can be the factor for stabilizing the M2 phase.

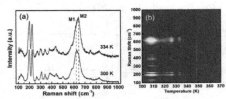

Figure 4. (a) Raman spectrum of as prepared VO_2 thin films at 300K and 334K, respectively;
(b) Mapping of Raman spectrum of as prepared VO_2 thin films.

CONCLUSIONS

In conclusion, epitaxial V_2O_3 and VO_2 thin films were synthesized on Al_2O_3 (0001) substrate by a polymer-assisted deposition technique. Both films exhibited metal insulator transition with sharp changes of resistance. The intermediate monoclinic M2 phase was observed during the MIT process by using Raman spectra-temperature mapping. The results indicated that PAD is a feasible way to grow high quality vanadium oxides thin films with controllable phases and desired MIT properties.

REFERENCE

1. F. Rodolakis, P. Hansmann, J. P. Rueff, A. Toschi, M. W. Haverkort, G. Sangiovanni, A. Tanaka, T. Saha-Dasgupta, O. K. Andersen, K. Held, M. Sikora, I. Alliot, J. P. Itie, F. Baudelet, P.

Wzietek, P. Metcalf and M. Marsi, Phys. Rev. Lett. **104**, 047401 (2010).

2. M. D. Goldflam, T. Driscoll, B. Chapler, O. Khatib, N. M. Jokerst, S. Palit, D. R. Smith, B. J. Kim, G. Seo, H. T. Kim, M. Di Ventra and D. N. Basov, Appl. Phys. Lett. **99**, 044103 (2010).

3. Z. L. Huang, S. H. Chen, C. H. Lv, Y. Huang and J. J. Lai, Appl. Phys. Lett.**101**, 191905 (2012).

4. C. Z. Yan, Z. Chen, Y. T. Peng, L. Guo and Y. F. Lu, Nanotechnology **23**, 475701 (2012).

5. H. Guo, K. Chen, Y. Oh, K. Wang, C. Dejoie, S. A. S. Asif, O. L. Warren, Z. W. Shan, J. Wu and A. M. Minor, Nano Lett. **11**, 3207 (2011).

6. C. Weber, D. D. O'Regan, N. D. M. Hine, M. C. Payne, G. Kotliar and P. B. Littlewood, Phys. Rev. Lett. **108**, 256402 (2012).

7. G. F. Zou, J. Zhao, H. M. Luo, T. M. McCleskey, A. K. Burrell and Q. X. Jia, Chem. Soc. Rev. **42** , 439 (2013).

8. B. J. Gao, F. Q. An and K. K. Liu, Appl. Sur. Sci. **253**, 1946 (2006).

9. Y. D. Ji, T. S. Pan, Z. Bi, W. Z. Liang, Y. Zhang, H. Z. Zeng, Q. Y. Wen, H. W. Zhang, C. L. Chen, Q. X. Jia and Y. Lin, Appl. Phys. Lett. **101**, 071902 (2012).

Mater. Res. Soc. Symp. Proc. Vol. 1547 © 2013 Materials Research Society
DOI: 10.1557/opl.2013.659

Effect of Adding Zn in $Cd_{1-x}Zn_xS$ Thin Films Prepared by an Ammonia-Free Chemical Bath Deposition Process

Iyali Carreón-Moncada[1], Luis A. González[1], Martin I. Pech-Canul[1], Rafael Ramírez-Bon[2]
[1]Centro de Investigación y Estudios Avanzados del IPN, Unidad Saltillo, Avenida Industrial Metalúrgica 1062, Parque industrial, Ramos Arizpe, CP.25900, Coah. México
[2]Centro de Investigación y de Estudios Avanzados del IPN, Unidad Querétaro, Apartado Postal 1-798, CP. 76001, Querétaro, Qro., México

ABSTRACT

The present investigation work shows the results of $Cd_{1-x}Zn_xS$ thin films (where X= 0.04, 0.08, 0.12, 0.16 and 0.2), obtained by total ammonia-free chemical bath processes. The reaction solutions were prepared with precursors of metallic salts as $CdCl_2$ and $ZnCl_2$ and replacing the ammonia with trisodic citrate ($C_6H_5O_7Na_3$) as complexing agent. The reaction solutions were stabilized with KOH to get alkaline solutions. As result of adding Zn, the as deposited films showed changes in their morphological, structural and optical properties. Moreover, additional changes were obtained when thermal treatments to 400°C under N_2 environment were applied to the as deposited films. The agglomerates at the surface of the annealed films showed larger grain sizes compared to that of the as deposited films. Due to preferential orientation of the hexagonal wurtzite-type structure in the films, changes in the intensity in the (002), (100) and (101) peaks from x-ray diffraction analysis were observed. Finally, a reduction on the maximum energy band gap from 2.65 to 2.59 eV was obtained as effect of the annealing treatment to the films.

INTRODUCTION

The $Cd_{1-x}Zn_xS$ is a ternary semiconductor mainly used as window material in the form of thin film for the manufacture and production of solar cells. This material provides the possibility of variations in its final composition, to obtain for example, a tuning in the band gap energy value [1-3]. Depending on the mechanism of deposition, the metallic ions of Cd and Zn are bonded to the sulfur to form binary compounds by separate or a mixture of them in the film. Therefore, these compounds may show the crystalline structures: zinc blend, hexagonal wurtzite or a combination of both. If an annealing treatment is applied to these films, often the remaining structure is the hexagonal wurtzite. This causes additional changes in the film properties, such as an enlargement of the agglomerates that form the film and modifications to their optical properties [4]. Some techniques for obtaining $Cd_{1-x}Zn_xS$ films are the Chemical Vapor Deposition (CVD), spray pyrolysis, Successive Ionic Adsorption and Reaction Layer (SILAR) and Chemical Bath Deposition (CBD). CBD is a simple, efficient and economical technique that has proved to be one of the bests for obtaining films of this compound with good homogeneity and uniformity [5-8]. Ammonia is a substance commonly used in the CBD technique for the deposition of sulfur compounds [9-16], due to its good performance as complexing agent. T. P. Kumar et al used a CBD process with ammonia to obtain thin films of $Cd_{1-x}Zn_xS$. According to the XRD analysis, the resulting films showed polycrystalline form and hexagonal structure with preferential orientation in the (0 0 2) direction. Energy band gaps of the $Cd_{1-x}Zn_xS$ samples obtained using optical absorption spectra varied between 2.27 and 3.25 eV [11]. However, ammonia is considered as an agent that affects the human health and the environment because of its toxicity. Due to its high volatility, ammonia produces changes in the deposition process that can lead to a lack of reproducibility of the films. To overcome this inconveniences, some proposals with ammonia free CBD have been developed [1, 17, 18]. Ammonia free CBD of $Cd_{1-x}Zn_xS$ thin films have not been yet reported. Based on our previous experience on the deposition of CdS films, here, we show results on the deposition of $Cd_{1-x}Zn_xS$ thin films on glass substrates with a completely ammonia free process [19, 20]. The resulting films were used to analyze the effect of adding Zn^{2+} ion precursor into the reaction solution on the morphological, structural and optical properties of the as deposited films. Changes in the properties of the films after annealing treatments to 400°C under N_2 environment are also analyzed and presented.

EXPERIMENT

The depositions of the films were done on glass substrates cleaned with detergent and then degreased with acetone, ethanol and distilled water into an ultrasonic cleaner. In order to obtain $Cd_{1-x}Zn_xS$ films, we prepared totally ammonia-free reaction solutions in beakers, using $CdCl_2$ 0.05M and $ZnCl_2$ 0.05M as metallic precursors, $CS(NH_2)_2$ 0.5M as source of S^{2+}ions and 0.5M $(C_6H_5O_7Na_3)$ as the complexing agent. In this case, proportions of precursors in solution X=0.04, 0.08, 0.12, 0.16 and 0.2 where $X=ZnCl_2/[CdCl_2+ZnCl_2]$ were used. The resulting solution was stabilized with KOH to pH=11.5. The temperature of deposition was 60°C. A study on the properties of the as deposited films was performed based on the results of the characterization with Scanning Electron Microscopy (SEM), Energy Dispersive Spectroscopy (EDS), X-Ray Difraction technique (XRD) and the UV-Vis Spectroscopy. Subsequently, the as deposited thin films were annealed at 400°C under N_2 environment, to study the changes on the surface morphology, elemental composition, crystalline structure and optical properties.

DISCUSSION

The color appearance of the as deposited films was changing gradually from orange to bright yellow with the increase of the mixture ratio X. According to the surface morphology analysis, the as deposited films showed agglomerates whose mean size increased with the mixture ratio X from 167.1nm to 405nm. In comparison, after the annealing treatments to 400°C, the agglomerates on the surface of the films were densified taking a spherical-like shape with larger grain size up to 568.71nm. This is illustrated in Fig. 1(a) and (b) for the as deposited and annealed films, respectively, obtained with X=0.2. In this case, the spectrum from EDS to identify the elemental composition of the films shows the content of Cd and S, but also the content of Zn whose intensity signals are visualized at 1 and 8.5 keV. The specific content of Cd, S and Zn for the as deposited film was 47.55at% of Cd, 12.98at% of Zn and 39.47at% of S, respectively. Meanwhile, that of the annealed film was 51.12at% of Cd, 5.70at% of Zn and 43.18at% of S. In table 1, we show a comparison of the elemental compositions obtained from EDS analysis for all of the films.

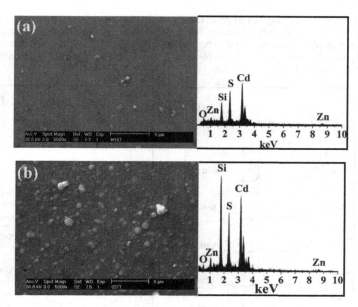

Figure 1. (right) Micrographs showing the form and size of the particles deposited in the $Cd_{1-x}Zn_xS$ thin films for X=0.2 and (left) corresponding EDS spectrum: (a) as deposited film and (b) annealed film.

Table 1. Elemental compositions of $Cd_{1-x}Zn_xS$ films

X	As deposited films (at%)			Annealed films (at%)		
	Cd	Zn	S	Cd	Zn	S
0.04	48.66	6.58	44.76	51.42	6.07	42.51
0.08	48.93	7.03	44.04	47.83	8.54	43.63
0.12	45.92	9.02	45.06	44.13	11.98	43.89
0.16	49.42	9.76	40.80	46.33	11.98	41.69
0.2	47.55	12.98	39.47	51.12	5.70	43.18

Fig. 2(a) shows the XRD pattern of the as deposited $Cd_{1-x}Zn_xS$ films showing three main peaks, which correspond to (100), (002), (101) reflections of the CdS wurzite-type hexagonal structure (PDF#00-041-1049). Previous experience on the deposition of CdS with similar ammonia-free formulation, indicates that a preferential growth orientation defined by the (002) crystalline direction is usually obtained [19, 20]. Here, note that the preferential growth orientation for all of the films changed to the (100) crystalline direction. This is consequence of stress by value differences of the ionic radii between Cd^{2+} (0.92 Å) and Zn^{2+} (0.60 Å). When the annealing treatment is applied to the films, in some cases there is a significant modification on the crystalline structure, as it is illustrated in Fig. 2(b). Specifically, the annealed film obtained with X=0.04 showed a change on its preferential growth orientation from the crystalline direction identified with the peak (100) to that of (002). Meanwhile, the film obtained with X = 0.12 showed an amorphous structure. Even when the intensity of the peak (100) was reduced for the annealed films obtained with X=0.08, 0.16 and 0.2, this crystalline direction still remains as the preferential growth orientation.

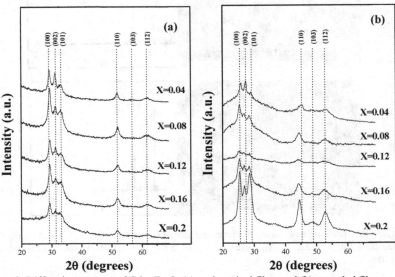

Figure 2. Diffraction patterns of Cd$_{1-x}$Zn$_x$S: (a) as deposited films and (b) annealed films

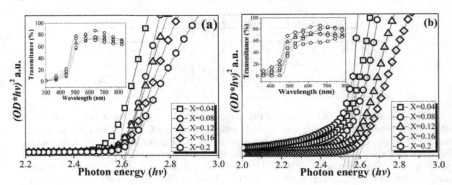

Figure 3. The (Optical density x hv)^2 versus energy spectrum, where the linear fit provides the energy band gap of the Cd$_{1-x}$Zn$_x$S films. The inset shows the optical transmittance: (a) as deposited films and (b) annealed films.

The analysis by UV-VIS spectrometry showed that the mean optical transmittance of 70%, for most of the films, showed no significant changes after the annealing process. However, for the case of X = 0.2, a decrease in the distribution of the optical transmittance of about 10% was observed after the annealing treatment. This is illustrated in Fig. 3(b).

Based on the optical transmittances, we calculated the direct bandgap values for the $Cd_{1-x}Zn_xS$ films [21]. In Fig. 3(a), it is observed a shifting of the bandgap values from 2.58eV to 2.65eV, for the as deposited films as effect of adding Zn. However, for all the cases the band gap values were shifted down when the films were annealed. Table 2 shows the resulting optical bandgap values for the as deposited films as well as that after the annealing treatment.

Table 2. Bandgap values calculated for as deposited and annealed $Cd_{1-x}Zn_xS$ films

X	As deposited films Eg (eV)	Annealed films Eg (eV)
0.04	2.58	2.55
0.08	2.62	2.56
0.12	2.62	2.57
0.16	2.63	2.59
0.2	2.65	2.56

CONCLUSIONS

$Cd_{1-x}Zn_xS$ thin films in an ammonia-free CBD process were obtained. The as deposited films showed important changes in their properties as effect of adding Zn. One of these effects was the increasing in the size of their agglomerates. The annealing treatments to these films produced an additional increase on the size of the agglomerates. The elemental compositions of the films showed no significant changes even after the annealing treatment for most of the films. However, there were noticeable changes in the crystalline structure for the annealed films according to the XRD analysis. The variation on the preferential growth orientation is result of a segregation process, in which, binary compounds of CdS and ZnS are formed. But, in other cases ternary compounds of $Cd_{1-x}Zn_xS$ can be achieved. The above changes also influenced on a shift of the bandgap to lower values. The above results, suggest us the use of these films on a variety of technological applications. For example, these films can be used as window materials with the capability of collecting photons from the blue portion of the solar spectrum.

ACKNOWLEDGMENTS

This work was supported by CONACYT through the becas program and project with reference number CB-2009-01-134572.

REFERENCES

1. G. Hodes. *Chemical Solution Deposition of Semiconductor Films*, 1st. ed. (Marcel Dekker, New York, 2003). p.280.
2. M. B. Ortuño. PhD. Tesis. Centro de investigación y de Estudios Avanzados del IPN, Unidad Queretaro. 2004.
3. A. Khare and S. Bhushan. Cryst. Res. Technol. 4, 7 (2006).
4. N.A. Noor, N. Ikrama, S. Alia, S. Nazirb, S.M. Alay-e-Abbasc, A. Shaukat. J. Alloys Compd. *2010*, 507.
5. G. Bruckman. Kolloid – z. 65, 1 (1933).
6. G. A. Kitaev, A. A. Uritskaya, S. G. Moksushin. Russ. J. Phys. Chem. *1965*, 39.
7. J. Kessler, J. Wennerberg, M. Bodegard, L. Stolt. Sol. Energy Mater. Sol. Cells. *2003*, 75
8. B. M. Basol, V. K. Kapur, C. R. Leidholm, A. Halan. Conference Record of the Twenty – Fifth IEEE Photovoltaic Specialists Conference (NREL Report No. TP - 410 - 21091). (1996) p.p. 157 – 162.
9. R. Mariappan, M. Ragavendar, V. Ponnuswamy. J. Alloys Compd. *2011*, 509.
10. S.D. Chavhan, S. Senthilarasu, L. Soo-Hyoung. Appl. Surf. Sci. *2008*, 254.
11. T. Prem Kumar, S. Saravanakumar, K. Sankaranarayanan, App. Surf. Sci. *2011*, 257.
12. T. Yamaguchi, Y. Yamamoto, T. Tanaka, Y. Demizu, A.Yoshida. Thin solid films. *1996*, 281-282.
13. V. B. Sanap, B. H. Pawar. Journal of Optoelectronics and Biomedical Materials, 2 (3), 39-43 (2011).
14. T. Prem Kumar, P. Ramesh, B.J. Anaraj. Chalcogenide lett. 8, 3 (2011).
15. D.C. Harris. *Quantitative Chemical Analysis*, 6th ed. (W.H. FREEMAN AND COMPANY, New York, 2003) p. 268-272.
16. A.G. Sharpe and C.E. Housecroft. *Inorganic Chemestry*, 2nd ed (Pearson Education Limited, Edinburgh, 2005), p. 229-232.
17. David B. Mitzi. *Solution Processing of Inorganic Materials*. (WILEY, New Jersey, 2009), p. 199.
18. S.M. Pawar, B.S. Pawar, J.H. Kim, J. Oh-Shim, C.D. Lokhande. Current Applied Physics. 2011
19. M.B. Ortuño-López, M. Sotelo-Lerma, A. Mendoza-Galván, R. Ramírez-Bon. Thin Solid Films. *2004*, 457.
20. G. Arreola-Jardon, L.A. González, L.A. García-Cerda, B. Gnade, M.A. Quevedo-Lopez, R. Ramírez-Bon. Thin Solid Films, *2010*, 519.
21. J. H. Simmons, *Optical materials*, K.S. Potter, Ed. Academic Press, USA (2000), p.p. 24 y 157

Mater. Res. Soc. Symp. Proc. Vol. 1547 © 2013 Materials Research Society
DOI: 10.1557/opl.2013.508

Fabrication of Flexible and Conductive Graphene-Silver Films by Polymer Dispersion and Coating Method

Jiang-Jen (JJ) Lin*, Sheng-Yen Shen, Po-Ta Shih and Rui-Xuan Dong
Institute of Polymer Science and Engineering, National Taiwan University, Taipei 10617, Taiwan.

ABSTRACT

We like to report a novel conductive film containing graphene-silver nanohybrids from the process of solution coating and annealing at low-temperature for melting silver nanoparticles (AgNPs) into interconnected Ag matrice on surface. The fabrication required the assistance of a home-made polymeric dispersant, poly(oxyethylene)-segmented imide (POE-imide), for homogenize the AgNPs and graphene in hybridized form. The intermediate dispersion of AgNPs at 10–25 nm diameter on the surface of 2D-graphene were characterized and subsequently subjected to solution coating into thin films. Under the annealing temperature as low as 160 °C, the films exhibited a high electric conductivity or low sheet resistance at 2.4×10^{-1} Ω/sq (equivalent to 7.9×10^4 S/cm). It is noteworthy that the significant point of low-temperature annealing at 160–170 °C that is attributed to the fast deterioration and degradation of the POE-imide organics kinetically before the AgNP coalescence and melting. Furthermore, the comparisons of using silicate clays and carbon nanotubes in replacing the 2D graphene for hybridizing Ag had revealed the different morphologies in Ag networks. The findings of using the polymeric dispersion for synthesizing nanohybrids may open up a new avenue for making films with integrated properties of flexibility, transparency and high conductivity for a host of electronic applications.

INTRODUCTION

In literature, graphene has drawn much attention due to its excellent electrical, thermal, and mechanical properties and its broad potential applications.[1,2,3] It is known that graphene sheets have high specific surface area which tended to form aggregation because of the van der Waals interaction. In literature, there are many methods to prepare the exfoliated graphene from graphite, such as chemical vapor deposition (CVD)[4], micromechanical exfoliation[5] and redox reaction[6]. However, graphene prepared by these methods has a common disadvantage of difficult dispersion in solution and in processing. Hence, surface modification of graphene is important research subject.[7]

On the other hands, Ag nanoparticles have the potential uses as electrical conductors besides their catalytic and antimicrobial properties.[8,9] Silver is known for using in microelectronics industries because of the highest electrical conductivity (6.3×10^5 S/cm) superior to most of other metals. For the practical applications, it is difficult to fabricate silver-coated polymer films or silver-polymer composite film with a low electrical resistance (<0.1 Ω/sq) under the process temperature of 300 °C.[10,11] Conventional methods indicate the uses of graphene oxide by acidification of graphite, then involving the reduction of AgNO$_3$ may effectively generate a conductive film.[12] In this work, we first use a home-made polymeric dispersant, poly(oxyethylene)-segmented oligo(imide) with multiple moieties of $-(CH_2CH_2O)_x-$ and imide $-(CONCO)-$ linking functionalities, for homogenizing the graphene and AgNP hybrid in solution medium. In the overall process via solution coating and annealing, a facile method was developed for preparing Ag/graphene hybrid film with high conductivity by low temperature heating.

EXPERIMENTAL DETAILS

Poly(oxyethylene)-segmented imide (POE-imide) dispersant was prepared from the oligomerization of poly(oxyethylene)-diamine and 4,4'-oxydiphthalic anhydride. The product mixture appeared to be yellowish waxy solid.

The nanohybrids of AgNPs on graphene surface were synthesized by *in situ* reduction of $AgNO_3$ in the presence of polymeric dispersion and graphene. Graphene (0.025 g) was first dispersed in 5 ml of DMF in a vial and sonicated under a VCX 500 Ultrasonicator at ambient temperature for 1 h. The resultant solution was dark black with some solid precipitates at the bottom of the container, indicating a low degree of dispersion. In a separate glass container, POE-imide (0.5 g) and $AgNO_3$ (0.5 g, 0.003 mole) were dissolved in 5 ml of deionized water (R = 18.2 M Ω/cm^2) and added to the graphene/DMF (5.0 ml) solution. A homogenous suspension of graphene was obtained by simple mixing in the presence of POE-imide and $AgNO_3$.

Conductive films were prepared by using a casting method. Typically, the matrix could be a rigid (e.g., glass) or a flexible (e.g., polyimide (PI)) film depending on the field of applications. The PI film (2 cm × 2 cm, see Figure 1) was washed several times with ethanol and used as the substrate. The graphene/Ag dispersion (0.5 g) was solution casted on a piece of PI film and was then placed in an oven with the heating program set as follows: 110, 160, 170, 300, and 350 °C for 1h of each temperature. Finally, flexible and conductive films were fabricated and characterized.

DISCUSSION

Preparation of POE-imide Polymeric Dispersant

Poly(oxyethylene)-segmented imide (POE-imide) was prepared by the imidization of poly(oxyethylene)-diamine and 4, 4'-oxydiphthalic anhydride in a molar ratio of 6:5 at the temperature of 180 °C (Scheme 1). The polymeric dispersant consisting of alternating POE – $(CH_2CH_2O)_x$– segment and aromatic imide –(CONCO)– linkages that promote the exfoliation of the graphene aggregates into delaminated nanosheets (Figure 1). The graphene nanosheet dispersion was used to in-situ reduce AgNO3 into AgNPs as the homogeneous nanohybrid. It was observed that AgNPs with a diameter distribution of 10–25 nm were located on the graphene surface through the non-covalent van der Waals force while some of AgNPs were freely remained in the solution (Figure 2). The in situ synthesis and performance of nanohybrids were also compared with the preparation of mixing of a homogeneous AgNPs and graphene that are individually prepared.

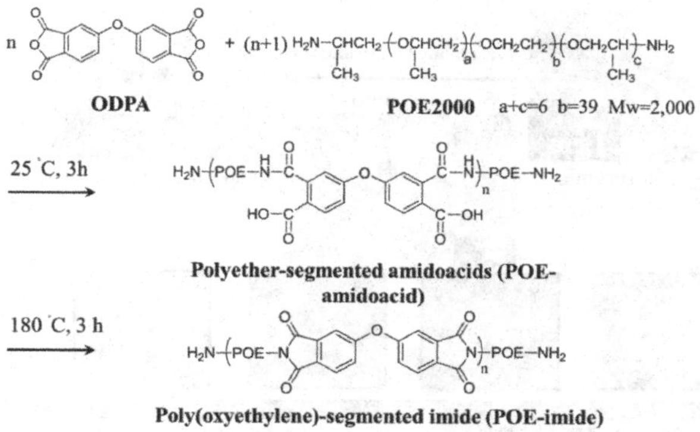

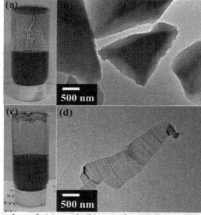

Scheme 1. Synthetic scheme for the poly(oxyethylene)- or poly(oxypropylene)-segmented amidoacid and imide as the dispersant.

Figure 1. TEM micrographs of (a) and (b) graphene dispersion (sonication), (c) and (d) graphene/POE-imide dispersion

Preparation of Graphene/POE-imide/Ag Films

The conductive films were fabricated by drop coating method according to procedures described as above. During the annealing at 150 °C ~ 160 °C, the film appeared to change color from red to golden (Figure 2). Further heating to 350 °C, the color changed into milky white. The corresponding measurements on electric resistance at the lowest 2.4×10^{-2} Ω/sq are shown in Table 1. When CNT or Mica was used to replace 2-D grapheme as the support matrices for AgNPs, all of the three matrices with AgNPs and POE-imide that conductive film will coalescence and melt at 300 °C, then reach a resistance as low as 1.2×10^{0} Ω/sq (Figure 3).

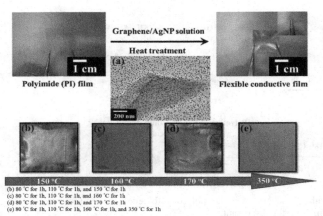

(b) 80 °C for 1h, 110 °C for 1h, and 150 °C for 1h
(c) 80 °C for 1h, 110 °C for 1h, and 160 °C for 1h
(d) 80 °C for 1h, 110 °C for 1h, and 170 °C for 1h
(e) 80 °C for 1h, 110 °C for 1h, 160 °C for 1h, and 350 °C for 1h

Figure 2. TEM of (a)AgNPs solution and photographs of melting AgNPs on the surface of polyimide film during the heating treatment. (b) 150 °C after 1h, (c) 160 °C after 1h, (d) 170 °C after 1h, and (d) 350 °C after 1h.

Table 1. Sheet resistance of the films prepared from the graphene/POE-imide/Ag, CNT/ POE-imide/Ag and Mica/POE-imide/Ag nanohybrids.

Sample[a]	Weight fraction (w/w/w)	Resistance (Ohm/sq)[b]				
		150 °C	160 °C	170 °C	300 °C	350 °C
CNT/AgNO$_3$/POE-imide[c]	1/20/20	——	2.1×10^5	2.0×10^{-1}	1.1×10^{-1}	1.0×10^{-2}
Mica/AgNO$_3$/POE-imide	1/20/20	2.3×10^7			1.2×10^0	
AgNO$_3$/POE-imide	1/1	——[d]	1.0×10^5	1.5×10^3	9.3×10^1	3.2×10^{-1}
Pristine Graphene(CPC)	1	1.4×10^3	1.4×10^3	1.4×10^3	1.4×10^3	1.4×10^3
Graphene/POE-imide	1/10	——	——	——	6.6×10^3	3.4×10^3
	1/20	——	——	——	8.1×10^3	5.7×10^3
Graphene /AgNO$_3$/POE-imide	1/10/10	0.6×10^5	4.3×10^0	1.1×10^0	8.4×10^{-1}	5.5×10^{-1}
	1/20/20	3.6×10^7	2.4×10^{-1}	1.2×10^{-1}	9.6×10^{-2}	2.4×10^{-2}

[a] solution drop coating on glass

[b] measured by four-point probe

[c] J. J. Lin et al. *ACS Appl. Mater. Interfaces*, **2012**, *4*, 1449-1455

[d] —— the resistance is too high to detect

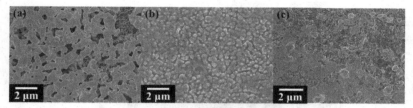

Figure 3. FE-SEM micrographs of (a) CNT/POE-imide/Ag, annealed at 350 °C; (b) graphene/POE-imide/Ag, annealed at 300 °C; (c) Mica/POE-imide/Ag, annealed at 300 °C on polyimide substrate.

Surface Morphologies of Graphene/Ag films

From the FE-SEM micrographs (Figure 4), we observed the AgNPs migrated and aggregated at 160 °C, and melted at 350 °C. The conductive films were demonstrated to serve as the connector for lighting up LED lamps (Figure 5), depending on the film compositions and annealing temperatures.

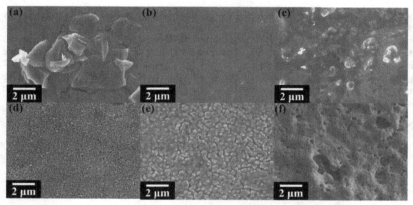

Figure 4. FE-SEM micrographs of (a) pristine graphene; (b) graphene:POE-imide hybrids with a weight ratio of 1:20, dried at 160 °C; (c) graphene:AgNO₃:POE-imide hybrids with a weight ratio of 1:20:20, annealed at 150 °C; (d) 160 °C; (e)170 °C and (f) 350 °C on polyimide substrate.

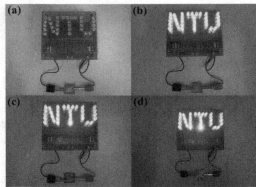

Figure 5. Demonstration of connecting electricity by the graphene/Ag films, annealed at (a) 150 °C, (b) 160 °C, (c) 170 °C and (d) 350 °C, respectively.

CONCLUSIONS

The POE-imide dispersant was essential for exfoliating graphene and then homogeneously dispersing AgNPs as the nanohybrids in a DMF/water medium. The AgNPs were tightly associated on graphene and free AgNPs were uniformly dispersed in medium with a diameter distribution of 10–25 nm. After being drop-casted on a PI substrate and annealed, a flexible and electrically conductive film was prepared and showed a high surface conductivity up to 2.4×10^{-1} Ω/sq, at the film annealing temperature of 160 °C and 9.6×10^{-2} Ω/sq at 300 °C. In the presence of graphene, AgNPs migrated and coalesced to generate a surface resistance with the result superior to the uses of CNT or clay as the supports.

ACKNOWLEDGEMENTS

We acknowledge financial supports from the Ministry of Economic Affairs (101-EC-17-A-08 -S1-205) and National Science Council (NSC) of Taiwan.

REFERENCES

1. W. Zhang, W. He, X. Jing, *J. Phys. Chem. B.*, **114**, 32(2010).
2. S. Park and R. S. Ruoff, *Nat. Nanotechnol.*, **4**, 217(2009).
3. Y. L. Huang, A. Baji, H. W. Tien, Y. K. Yang, S. Y. Yang, S. Y. Wu, C. C. Ma, H. Y. Liu, Y. W. Mai, N. H. Wang, *Carbon*, **50**, 3473(2012).
4. T. Aizawa, R. Souda, S. Otani, Y. Ishizawa, C. Oshima. *Phys. Rev. Lett.*, **64**, 768(1990).
5. K. S. Novoselov, A. K. Geim, S. V. Morozov, D. Jiang, Y. Zhang, S. V. Dubonos,et al. *Science*, **306**, 666(2004).
6. S. Sasha, R. D. Piner, X. Chen, N. Wu, S. T. Nguyen and R. S. Ruoff, *J. Mater. Chem.*, **16**, 155(2006).
7. T. M. Swager, *ACS Macro Lett.*, **1**, 3(2012).
8. Y. Oh, K. Y. Chun, E. Lee, Y. J. Kim, S. Baik, *J. Mater. Chem.*, **20**, 3579(2010).

9. H. L. Su, C. C. Chou, D. J. Hung, S. H. Lin, I. C. Pao, J. H. Lin, F. L. Huang, R. X. Dong, J. J. Lin, *Biomaterials*, **30**, 5979(2009).
10. S. Qi, Z. Wu, D. Wu, W. Wang, R Jin, *Langmuir*, **23**,4878(2007).
11. D. S. Thompson, L. M. Davis, D. W. Thompson, and R. E. Southward, *ACS Appl. Mater. Interfaces.*,**1**,1457(2009).
12. H. W. Tien, Y. L. Huang, S. Y. Yang, J. Y. Wang, C. C. Ma, *Carbon.*, **49**,1550(2011).
13. R. X. Dong, C.T. Liu, K.C. Huang, W. Y. Chiu, K. C. Ho, J. J. Lin, *ACS Appl. Mater. Interfaces.*,**4**,1449(2012).
14. C. W. Chiu, P. D. Hong and J. J. Lin, *Langmuir*, **27**, 11690(2011).

Ferroelectrics and Multiferroics

Mater. Res. Soc. Symp. Proc. Vol. 1547 © 2013 Materials Research Society
DOI: 10.1557/opl.2013.634

Hydrothermal epitaxy of lead free (Na,K)NbO₃-based piezoelectric films

Albertus D. Handoko[1] and Gregory K. L. Goh[1,*]
[1]Institute of Materials Research and Engineering, 3 Research Link,
Singapore 117602, Singapore.

ABSTRACT

Lead free niobate solid solutions can exhibit piezoelectric properties comparable to that of lead zirconate titanate piezoelectrics in the vicinity of its morphotropic phase boundary (MPB). Here we describe how $(Na,K)NbO_3$ and $(Na,K)NbO_3$-$LiTaO_3$ solid solution thin films can be grown epitaxially by the hydrothermal method at temperatures of 200 °C or below in water and be made ferro- and piezoelectrically active by a simple 2 step post growth treatment.

INTRODUCTION

It has been shown[1,2] that lead free $(Na,K)NbO_3$ based (NKN) solid solutions can exhibit piezoelectric properties comparable to that of lead zirconate titanate (PZT) piezoelectrics in the vicinity of its morphotropic phase boundary (MPB). Unfortunately current techniques available to grow NKN based solid solution films require high temperatures, sometimes in excess of 1000 °C, which are not only detrimental to the stoichiometry due to volatility of alkali components[3] but also cause unwanted thermal-related defects such as twins and cracking upon cooling through the Curie temperature.

Hydrothermal synthesis is a novel and unique materials synthesis route that allows the formation of various compounds including epitaxial films at mild conditions, using water as the main solvent.[4] Advantageously, this method is not only simple but also self contained and easily scalable. Waste recovery and solution reutilisation is also possible via simple centrifugation process.[5] Additionally, films grown hydrothermally can virtually be free of thermal defects, because the growth temperature is typically below most materials Curie transition temperature,[4] although proton/water incorporation remains a great challenge for the electrical properties which needs to be subsequently removed by post-growth treatments at modest conditions.[6,7]

In this study, hydrothermal methods are explored to grow single-phase, complex composition $(Na,K)NbO_3$-$LiTaO_3$ (NKN-LT) solid solution thin films. Ultimately we are seeking compositions as close to the MPB as possible to obtain reasonably high polarisation and piezoelectric constant (d_{33}) values in order to be considered as viable lead-free alternatives to $Pb(Zr,Ti)O_3$ based systems. Complications and challenges to grow NKN-LT films in a single-step, single-pot synthesis is also described.

EXPERIMENTAL

NKN powder was synthesized by reacting 3.76 mmol of Nb_2O_5 (99.99%, Aldrich, St. Louis, MO) in 25 ml aqueous mixtures of KOH (>88%, J.T. Baker, Phillipsburg, NJ) and NaOH (>98%, GCE) at 200 °C for 24 hours in a PTFE lined stainless steel autoclave (Parr Co., Moline, IL). The Nb_2O_5 precursor (99.99%, Aldrich, St. Louis, USA) was identified to be a mixture of

91% H-Nb$_2$O$_5$ [8] and 9% T-Nb$_2$O$_5$ [9]. In all cases, the total [OH$^-$] was fixed at 6 M because it was found to be the optimum concentration for the reaction to complete in 24 hours at 200 °C.

After the reaction was completed, the resulting powder and solution were centrifuged at 3500 rpm for 10 min. The remaining powder slurry was washed with deionised water (Millipore, MA, resistivity >18.2 MΩ•cm) at room temperature 4 to 5 times until the pH value was neutral as indicated by litmus paper. The white powder slurry was then dried at 60°C in air for one day, weighed and stored in a dry box prior to further analysis. To investigate the effects of post growth annealing on the as synthesized powder's phase stability, heat treatment was carried out in a platinum lined alumina crucible at 800 °C for 2 hours with a ramp rate of 10 °C min^{-1}.

Epitaxial NKN and NKN-LT films were grown using similar procedures with NKN powders but this time 10 by 5 mm (100)-oriented Nb-doped SrTiO$_3$ (Nb:STO) single crystal substrates (MTI Corp., Richmond, USA) were introduced face down 20 mm above the bottom of the PTFE-lined stainless steel hydrothermal reactor. The substrates were first cleaned in an ultrasonic bath in isopropanol and deionised (DI) water for 5 minutes each before the growth procedure. Similar to NKN powder growth, appropriate amounts of NaOH and KOH solutions were introduced based on the desired R ratio for a 25 ml total volume. For NKN-LT, small amounts of LiCl (99%, Sigma Aldrich, St. Louis, USA) were also introduced together with the NaOH and KOH mixture. The total [OH$^-$] was also kept at the optimal 6 M concentration. The NaOH-KOH-LiCl solution mixture was then transferred into the hydrothermal reactor containing 3.76 mmol of Nb$_2$O$_5$ or mixture of Nb$_2$O$_5$ and Ta$_2$O$_5$ (99.99%, Aldrich, St. Louis, USA).

The solid solution composition in both film and powders were varied by changing the value of R and T ratios from 0 to 20% where

$$R = \frac{[NaOH]}{[NaOH]+[KOH]}$$

$$T = \frac{[Ta_2O_5]}{[Ta_2O_5]+[Nb_2O_5]}$$

The total Li$^+$ ion concentration (in the form of LiCl) in this precursor was between 0.15 – 0.3 M. 0.1 g of ethylenediaminetetraaceticacid (EDTA, 99.8%, Sigma Aldrich, St. Louis, USA) was added to the 25 ml prepared film precursor solution to promote repeatable NKN-LT film growth.

As grown NKN-LT films were subjected to a post-growth treatment process comprised of O$_2$ plasma followed by thermal annealing reported previously to bring out the best electrical properties of hydrothermally grown KNbO$_3$ films.[10] The O$_2$ plasma treatment was carried out in a reactive ion etcher (Trion Tech., Sunnyvale, USA) set at 200 W with 400 mTorr O$_2$ pressure for 15 mins. Subsequent thermal annealing was done in a box furnace at 600 °C for 3 hours using a double crucible setting. The NKN-LT films were placed in an inner crucible, with 0.05 : 0.45 : 0.45 mixture of Li$_2$CO$_3$: Na$_2$CO$_3$: K$_2$CO$_3$ powders placed in the outer crucible to mitigate alkali oxide volatility.

Powder and film X-ray diffraction (XRD) was measured using a Bruker D8 diffractometer with a Xe gas-filled general area detector with copper radiation at 5 kV 10 mA, 300 seconds acquisition time which covered a two-dimensional area range of approximately 27 °2θ across and around 30 to 100 °χ each acquisition frame depending on the detector position. The two-dimensional area detector allows assessment of film texture and epitaxial growth.

PANalytical X'Pert PRO High Resolution X-ray Diffractometer (HRXRD) [Cu-$k\alpha$ radiation operating at 40 kV voltage, 40 mA current] was also used for higher resolution film diffraction to check for any crystalline impurity phases. Scanning electron micrographs (SEM) were obtained using a JEOL-6700 FESEM. Depth-resolved elemental analysis was done using a TOF-SIMS IV time-of-flight secondary ion mass spectrometer (ION-TOF GmbH, Münster, Germany) equipped with 25 keV Ga gun for analysis, covering approx. 100 x 100 μm^2 film surface area.

To investigate the electrical and ferroelectric properties, 1 mm diameter circular gold electrodes were sputtered onto the surface of the films, effectively setting up a parallel plate capacitor configuration with the conducting oxide substrate as the bottom electrode. Ferroelectric response of the films were tested using a standard Precision RT66A ferroelectric test unit (Radiant Technologies, Albuquerque, USA) run by Vision 4 software on a hysteresis set program. Piezoelectric response was measured by a laser scanning vibrometer comprised of OFV-056 scanning head and OFV-3001-SF6 vibrometer controller (Polytec GmbH, Waldbronn, Germany). The advantage of using a laser scanning vibrometry (LSV) compared to conventional interferometer system is that multiple points can be rastered within a single experiment, removing the need for equipment readjustment that could lead to measurement errors.[11] More importantly, LSV is able to detect both the vibration magnitude and phase of the film and substrate simultaneously, therefore the error caused by substrate bending or movement can be properly taken into account.[12]

RESULTS AND DISCUSSION

One of the major challenge to grow niobate solid solution films hydrothermally is the separation between the sodium-rich impurities and the desired NKN phase when precursor mixtures with higher R ratio (higher sodium content) is used. Fig. 1 shows typical phase separation in NKN powders where Na-rich impurity phase (denoted by the asterisks) starts appearing when R ratio is approx. 13% (upper trace) and dominates when the ratio is higher than 20% (lower trace). The main reason for the phase separation is the different solubility of the intermediate phase, of which sodium intermediates precipitates out first and quickly becomes Na-rich niobates.[13] This phase separation seen in powder growth also creates the same problem during film growth because the growth rate of Na-rich niobates are known be faster than potassium rich counterpart[14] and blocks growth of NKN films with the desired Na to K ratio near the MPB composition.

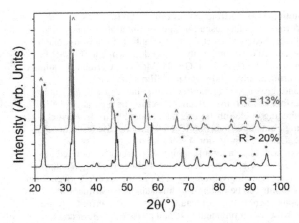

Figure 1: X-ray diffraction data of NKN powders at different R ratio of 13% (upper trace) and 20% (lower trace). Na-rich impurities (*) start to appear at R=13% and dominates above R=20. Desired NKN phase is marked with (^).

Indeed initial NKN film growth without any modification always result in two separate phases; the expected NKN phase reflexion at around 45 °2θ and an additional shoulder to the left of the substrate reflexion around 46 °2θ (Fig. 2a). The slightly larger 2θ position of the impurity phase is indicative of smaller lattice parameter of the high resolution, pointing towards Na-rich niobates. Compositional depth profiling using TOF-SIMS (Fig. 2b) revealed crucial information that the Na-rich impurities may actually be growing on the surface (i.e. at a later growth stage) rather near the interface with the substrate, indicated by the steep increase of Na concentration near the film surface. This information presents opportunity to obtain the desired phase pure NKN film by either applying post-growth thermal treatments to allow homogenisation of the Na/K ratio or trying to arrest the Na-rich phase growth during the hydrothermal process by parameter modification or introduction of growth inhibitors.

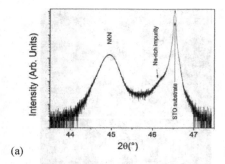 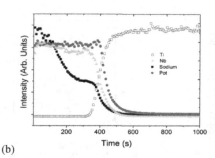

(a)   2θ(°)   (b)

Figure 2: (a) HRXRD of as grown NKN film (R ≈ 14.1%) without any additional growth agent, showing additional broad shoulder on the left of SrTiO3 substrate peak, indicative of Na-rich impurity phase with smaller lattice parameter. (b) Depth profiling via SIMS shows uneven sodium content (filled squares) across the film thickness, indicating that the Na-rich impurity grows later and accumulates on top of the film. Titanium signal (hollow squares) was monitored to gauge the film-substrate interface.

The information obtained from NKN film growth was then applied to make phase pure, more complex NKN-LT films. The additional lithium and tantalum dopant are responsible to create a further phase boundarie between orthorhombic NKN and trigonal LiTaO₃, enhancing its ferro-/piezoelectric properties.[1] The two-dimensional area detector XRD suggests that epitaxial phase pure NKN-LT films could be obtained after lowering the growth temperature to 130 °C and adding EDTA as Na-rich phase inhibitor (Fig. 3a-b). Absence of arc rings on the area detector confirms that the film was indeed epitaxial. No other reflexions were detected apart from substrate and NKN-LT film contribution. It was proposed that EDTA suppresses the Na-rich phase and slows down the overall growth rate such that pure NKN-LT films can grow unhindered. [15] The lowering of growth temperature is also crucial to reduce crystal growth, without which rapid subsequent growth of Na-rich impurities can still grow, leading to excessively thick films dominated by Na-rich phase. Subsequent composition depth profiling via TOF-SIMS measurement confirms the presence of all NKN-LT elemental constituents and shows that the relative proportion were reasonably constant across the film thickness.

The resulting NKN-LT film was also found to be sufficiently dense and smooth, with thickness of approx. 0.5 μm estimated from the cross sectional electron micrographs (not shown). This allows for good electrical contact and sufficient dielectric constant for subsequent electrical measurements.

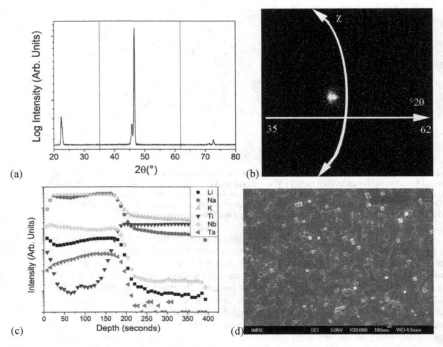

Figure 3: Characterisation of as-grown NKN-LT films: (a) the integrated intensities of area detector XRD, (b) actual two-dimensional spectra captured by the area detector, the 2θ range encompass the marked area of Fig. 3a, between 35 and 62 °2θ. (c) Composition depth profile taken by TOF-SIMS indicates that all NKN-LT components were accounted for and the relative proportions were reasonably constant. (d) FESEM micrograph of NKN-LT film, showing reasonably smooth and dense film, allowing for good electrical contact.

It has been known that hydrothermally grown film electrical properties suffer due to possible incorporation of protons,[7] and various post-growth treatment methods such as heat treatment[16] and oxygen plasma treatment[17] have been suggested. However, recently it was discovered that oxygen plasma and heat treatment combination is required to bring out the best properties in hydrothermally grown film.[10] Here, it is found that the same combination treatment is also effective in improving ferro-/piezoelectric response of NKN-LT films. As described in Fig. 4, the NKN-LT film can only be poled to saturation after the combination treatment (bigger loop). As a comparison, the as-grown NKN-LT films show leaky dielectric properties (not shown) and was not able to withstand greater than 50 kV/cm applied field (smaller loop). The treated NKN-LT films also show clear and reasonable piezoelectric response (Fig. 4b). Based on the ≈50 pm displacement recorded by the laser vibrometry technique for a 2 V bias applied to the film, a conservative estimate of its d_{33} piezoelectric constant was around 25 pm/V. No comparison could be made with as-grown films because no displacements were detected.

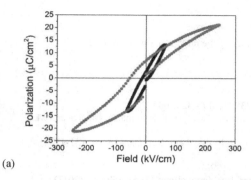

(a)

(b)

Figure 4: (a) Ferroelectric measurement of 15 minutes oxygen plasma- 600 °C 3 hrs heat treatment combination of NKN-LT film (R 16.7%, T=20%, LiCl = 0.15 M). Smaller loop is for as grown film (b) The actual displacement scanned by laser vibrometry technique, showing approx. 50-60 pm displacement.

CONCLUSIONS

$(Na,K)NbO_3$-$LiTaO_3$ based solid solution is one of the forefront lead-free ferro-/piezoelectric candidate to replace PZT based materials. The realisation of facile film growth at reasonably low temperature is essential if we were to see real implementation of lead-free ferro-/piezoelectric compounds in foreseeable future. Here we have deliberated that such growth is possible via hydrothermal method, a scalable process that utilises widely available precursors at low temperature of 130 °C in water. Subsequent oxygen plasma – heat treatment combination could then bring out the best piezoelectric constant, d_{33}, of 25 pm V^{-1}, as determined by laser scanning vibrometry. This value can still be greatly improved as the NKN-LT films composition is not yet close to the phase boundary.

ACKNOWLEDGMENTS

The authors thank Dr. Goh Phoi Chin for conducting the laser scanning vibrometer measurements.

REFERENCES

1.   Y. Saito and H. Takao, Ferroelectrics **338** (1), 17 - 32 (2006).
2.   Y. Saito, H. Takao, T. Tani, T. Nonoyama, K. Takatori, T. Homma, T. Nagaya and M. Nakamura, Nature **432** (7013), 84-87 (2004).
3.   R. H. Lamoreaux and D. L. Hildenbrand, J. Phys. Chem. Ref. Data **13** (1), 151-173 (1984).
4.   F. F. Lange, Science **273** (5277), 903-909 (1996).
5.   A. D. Handoko and G. K. L. Goh, Mater. Res. Innovations **15** (5), 352-356 (2011).
6.   C. K. Tan and G. K. L. Goh, Thin Solid Films **515** (16), 6572-6576 (2007).
7.   C. K. Tan, G. K. L. Goh and W. L. Cheah, Thin Solid Films **515** (16), 6577-6581 (2007).
8.   K. Kato, Acta Crystallogr., Sect. B **32** (3), 764-767 (1976).
9.   K. Kato and S. Tamura, Acta Crystallogr., Sect. B **31** (3), 673-677 (1975).
10. A. D. Handoko, G. K. L. Goh and R. X. Chew, CrystEngComm **14** (2), 421-427 (2012).
11. K. Yao, S. Shannigrahi and F. E. H. Tay, Sensor Actuat. A: Phys. **112** (1), 127-133 (2004).
12. K. Yao and F. E. H. Tay, Ultrasonics, Ferroelectrics and Frequency Control, IEEE Transactions on **50** (2), 113-116 (2003).
13. A. D. Handoko and G. K. L. Goh, Green Chem. **12** (4), 680-687 (2010).
14. G. K. L. Goh, F. F. Lange, S. M. Haile and C. G. Levi, J. Mater. Res. **18** (2), 338-345 (2003).
15. A. D. Handoko and G. K. L. Goh, CrystEngComm **15** (4), 672-678 (2013).
16. A. T. Chien, X. Xu, J. H. Kim, J. Sachleben, J. S. Speck and F. F. Lange, J. Mater. Res. **14** (8), 3330-3339 (1999).
17. I. R. Abothu, P. M. Raj, D. Balaraman, M. D. Sacks, S. Bhattacharya and R. R. Tummala, presented at the 9th International Symposium onAdvanced Packaging Materials: Processes, Properties and Interfaces, Atlanta, GA, 2004 (unpublished).

Mater. Res. Soc. Symp. Proc. Vol. 1547 © 2013 Materials Research Society
DOI: 10.1557/opl.2013.855

Field Dependent Carrier Transport Mechanisms in Metal-Insulator–Metal Devices with Ba$_{0.8}$Sr$_{0.2}$TiO$_3$/ ZrO$_2$ Heterostructured Thin Films as the Dielectric

Santosh K. Sahoo,[1,2] H. Bakhru,[3] Sumit Kumar,[4] D. Misra,[2] Colin A. Wolden,[5] Y. N. Mohapatra,[6] and D. C. Agrawal[6]

[1]National Renewable Energy Laboratory, 1617 Cole Blvd., Golden, CO 80401, USA

[2]New Jersey Institute of Technology, Newark, NJ 07102, USA

[3]College of Nanoscale Science and Engineering, SUNY Albany, NY 12222, USA

[4]Intel Corporation, 5000 W Chandler Blvd., Chandler, AZ 85226, USA

[5]Chemical and Biological Engineering, Colorado School of Mines, Golden, CO 80401, USA

[6]Materials Science Programme, Indian Institute of Technology, Kanpur 208016, India

ABSTRACT

Ba$_{0.8}$Sr$_{0.2}$TiO$_3$/ZrO$_2$ heterostructured thin films with different individual layer ZrO$_2$ thicknesses are deposited on Pt/Ti/SiO$_2$/Si substrates by a sol-gel process. The current versus voltage (I-V) measurements of the above multilayered thin films in metal-insulator-metal (MIM) device structures are taken in the temperature range of 310 to 410K. The electrical conduction mechanisms contributing to the leakage current at different field regions have been studied in this work. Various models are used to know the different conduction mechanisms responsible for the leakage current in these devices. It is observed that Poole-Frenkel mechanism is the dominant conduction process in the high field region with deep electron trap energy levels (φ_t) whereas space charge limited current (SCLC) mechanism is contributing to the leakage current in the medium field region with shallow electron trap levels (E_t). Also, it is seen that Ohmic conduction process is the dominant mechanism in the low field region having activation energy (E_a) for the electrons. The estimated trap level energy varies from 0.2 to 1.31 eV for deep level traps and from 0.08 to 0.18 eV for shallow level traps whereas the activation energy for electrons in ohmic conduction process varies from 0.05 to 0.17 eV with the increase of ZrO$_2$ sub layer thickness. An energy band diagram is given to explain the dominance of the various leakage mechanisms in different field regions for these heterostructured thin films.

INTRODUCTION

Nowadays, high K materials play an important role in microelectronic devices such as capacitors and memory devices [1]. Recently, ferroelectric barium strontium titanate [Ba$_x$Sr$_{1-x}$TiO$_3$, (BST)] thin film has attracted much attention for its potential applications in devices such as dynamic random access memories (DRAM), field effect transistors (FET) because of its high dielectric constant, low leakage current, and high breakdown field [2-4]. The leakage current of the BST thin film is one of the most important issues in estimating the charge retention capacity of the capacitor [4]. Also power dissipation in memory devices is due to the leakage current through the gate dielectric material [5]. Thus, to reduce the power dissipation and for obtaining a better charge retention of DRAM a low leakage current is essential. Among the various approaches to optimize the properties in BST thin films, recently, a multilayer structure with a combination of BST and other dielectrics has been proven to be useful approach to improve the properties of BST film. Various multilayer structures with BST as the main constituent have been studied such as BST/SiO$_2$/BST, BST/ZrO$_2$/BST, BST/MgO/BST, etc. [6-8] and different properties enhancement have also been reported. Among the different approaches to reduce the leakage current in BST thin film, a multilayer with a combination of BST and ZrO$_2$ has been proven to be very useful approach [7]. The conduction current is one of the most important issues for capacitor charge retention. In order to fully understand the conduction current, the carrier transport mechanisms should be studied thoroughly.

Leakage current analysis in the dielectric thin film has been an important study for the use of these materials in microelectronic devices. The conduction mechanisms in thin films can be classified into two groups such as interfaced-controlled including Schottky emission, Fowler-Nordheim (FN) tunneling and bulk-controlled including Poole-Frenkel emission, space- charge-limited conduction (SCLC), and Ohmic conduction process [9,10]. Lots of work has been done to reduce the leakage currents in the heterostructured BST/MgO and BST/ZrO$_2$ thin films [7,11]. However, the field dependent electrical conduction mechanism in pure BST films and BST/ZrO$_2$ heterostructured films and the study of trap level energies and activation energies have not been reported in detail yet.

In this work, the leakage current in BST films and BST/ZrO$_2$ multilayer thin films has been measured in the temperature range of 310 K to 410 K. The conduction mechanisms contributing to the leakage current in different field regions and trap energy levels as well as activation energies in these films have been studied based on the different models specific to the dielectric thin films.

EXPERIMENTAL

Ba$_{0.8}$Sr$_{0.2}$TiO$_3$/ZrO$_2$/Ba$_{0.8}$Sr$_{0.2}$TiO$_3$ heterostructured thin films are prepared on Pt/Ti/SiO$_2$/Si substrates by a sol-gel process. The details of the preparation of pure Ba$_{0.8}$Sr$_{0.2}$TiO$_3$ sol and ZrO$_2$ sol can be found elsewhere [7]. Three different types of films are prepared such as pure BST (B0), BST/ZrO$_2$/BST (1 ZrO$_2$ layer between) (BZ1) and BZ4 (4 ZrO$_2$ layers between) keeping the total thickness of the films as 380 nm as mentioned in our earlier work [7]. Top electrodes of Au, 0.5 mm in diameter, are deposited on thin films to make MIM devices for electrical measurements. The electroded film is annealed at 200°C for 5 minutes to make good contact between the film and the top electrode. A Keithley 6517A electrometer and a heating stage (MMR technology) are used for temperature dependent current density versus field (J-E) measurements in the range from 310 to 410 K.

RESULTS AND DISCUSSION

The XRD of the film (not shown) has peaks from BST and Pt substrate. There is an additional peak (111) at $2\theta = 30°$ due to cubic phase of ZrO$_2$. Figure 1 (a) shows the leakage current density (J) versus electric field (E) plots for the multilayered films (BZ4) in the temperature range of 310 - 410 K. It is observed that J increases with the increase of temperature. The strong temperature dependence of the leakage current density (Fig.1(a)) suggests that the electron transport is activated such as one would expect either a Schottky-barrier mechanism or Poole-Frenkel conduction is contributing to the leakage current in these multilayered films. It is known that the Schottky mechanism is due to the flow of electrons over a barrier at the metal-insulator interface. The conduction current due to Schottky mechanism is described using the following relation [9]

$$J_{SC} = AT^2 \exp\left(\frac{\beta_{SC} E^{1/2} - q\Phi_b}{kT}\right) \qquad (1)$$

where A is the effective Richardson constant, $q\Phi_b$ is the Schottky barrier height, and β_{SC} = $(q^3/4\pi\varepsilon\varepsilon_0)^{1/2}$ (ε is the dynamic dielectric constant and ε_0 is the permittivity of the free space). To evaluate the contribution of the Schottky mechanism to the leakage current the I-V data is plotted as $ln\ (J/T^2)$ versus $E^{0.5}$ (not shown). It is observed that a linear fit is obtained in the high field region for all the temperatures. The values of dynamic dielectric constant, ε, extracted from the slope of the linear fit in the Schottky plots comes out to be in the range of 0.75 to 0.17 which is unrealistically low compared to the expected value of $\varepsilon \sim n^2 = 4.0$ [10]. Therefore, Schottky mechanism is not contributing to the leakage current under these conditions.

Next, we explored the possibility that the Poole-Frenkel mechanism may be contributing to the leakage current. It is well known that the Poole-Frenkel mechanism is due to the field enhanced thermal excitation of trapped electrons into the conduction band. Defects below the conduction band within the dielectric bandgap can participate in the conduction process through electron trapping and detrapping when both field and temperature are varied. The leakage current density due to Poole-Frenkel mechanism can be described by the following equation [9]

$$J_{PF} = qN_c\mu E \exp\left(\frac{\beta_{PF} E^{1/2} - q\phi_t}{kT}\right) \qquad (2)$$

where $\beta_{PF} = (q^3/\pi\varepsilon\varepsilon_0)^{1/2}$ (ε and ε_0 are defined in Eq.1), N_c is the density of states in the conduction band, μ is the electron mobility, $q\phi_t$ is the trap level energy, E is the electric field, k is the Boltzmann constant, and T is the device temperature.

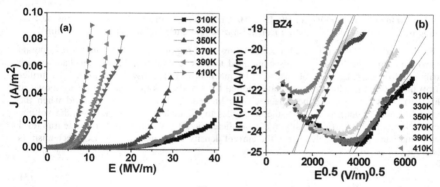

Figure 1. J vs. E plot (a) and ln (J/E) vs. $E^{0.5}$ (b) for BZ4 heterostructured thin films at different temperatures.

The potential contribution of the Poole-Frenkel mechanism is examined by plotting $ln\ (J/E)$ versus $E^{0.5}$ as shown in Fig. 1(b). It is observed that a straight line fit is obtained in the high field region for the temperatures in the range of 310 K to 410 K. The dynamic dielectric constant ε entering the current density versus field relation can be extracted from the slope of the linear fit. The refractive index, n, of BST film is about 2.0 [10] so that the expected value of $\varepsilon \sim n^2$ is about

4.0 [10]. The values of ε determined from the Poole-Frenkel plots come out to be in the range of 3.1 to 2.1 as the temperature increases from 310 to 410 K. Thus, the ε values obtained from PF plots are close to 4.0 with a better agreement compared to those obtained previously from Schottky plots. Therefore, the Poole-Frenkel mechanism is contributing to the leakage current in the high field region for the multilayered samples. According to equation 2, a plot of ln (J/E) vs. $E^{1/2}$ would yield a straight line (Fig. 1(b)) and additionally, when measured at constant electric field E, the plot of *ln (J/E)* vs. *1000/T* also would yield a straight line as depicted in Figure 2(a). The estimated trap energy level ($q\varphi_t$) from the slope of the linear fit in Figure 2(a) comes out to be in the range of 0.2 to 1.31 eV as the individual ZrO_2 layer thickness increases.

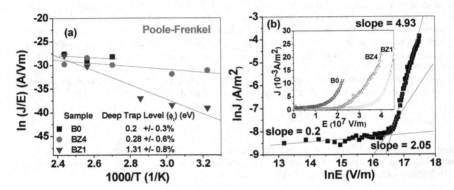

Figure 2. (a) ln (J/E) vs. 1000/T for B0, BZ1, BZ4 films and (b) ln J vs. ln E at 310K for BZ4 film and the inset shows J vs. E comparison at room temperature for B0, BZ1 and BZ4 films.

From figure 1(b), it is observed that in the low field region the leakage current data points are deviating from the linear fit. This concluded that some other conduction mechanisms are contributing to the leakage current instead of PF mechanism in the low field regimes. To further confirm the leakage mechanisms, we have plotted the I-V data as ln J versus ln E as shown in figure 2(b). It is well known that space charge limited current effect is observed when the injected free carrier density, n_i, from the electrodes to the dielectric is higher than the volume generated intrinsic free carrier density, n_0, in the dielectric [10]. When a single discrete trap level exists in the band gap of the dielectric, the current density J dominated by SCLC mechanism is typically given as [9]

$$J = \frac{9}{8} \mu \varepsilon_0 \varepsilon_r \theta \frac{E^2}{d} \tag{3}$$

where ε_r is the relative dielectric constant, ε_0 is the permittivity of free space, μ is the electron mobility, E is the applied electric field, d is the film thickness and θ is the ratio of free to trapped charge. According to equation (3), when ln J versus ln E is plotted (figure 2(b)) the slope of the straight line becomes ~ 2. Therefore, SCLC mechanism is the dominant conduction process in the medium field region for these multilayered films.

The SCLC mechanism was further examined through analysis of the trap energy levels. If the traps present in the dielectric film are shallow, θ can be given by [12]

$$\theta = \frac{N_C}{N_t} \exp\left[\frac{-(E_C - E_t)}{kT}\right] \quad (4)$$

where N_t is the trap density, N_C is the density of states in the conduction band, $(E_C - E_t)$ is the activation energy for the shallow electron traps, and T is the device temperature. Since N_C in θ in Eq. (4) is proportional to $T^{3/2}$ and the electron mobility μ in Eq. (3) is proportional to $T^{-1/2}$, therefore the temperature dependence of the current density J in the case of SCLC transport is given by [12]

$$J \propto T \exp\left[\frac{-(E_C - E_t)}{kT}\right] \quad (5)$$

To obtain the activation energy for the shallow electron traps for the SCLC conduction process, $\ln(J/T)$ versus $1000/T$ is plotted at a fixed electric field for BST films and BST/ZrO$_2$ multilayered films as shown in Fig. 3(a). The shallow trap level energy (E_t) estimated from the slope of the linear fits comes out to be in the range of 0.08 to 0.26 eV as the ZrO$_2$ layer thickness increases. The estimated shallow trap level energy of 0.08 eV for our BST film is close to the reported value of 0.14 eV [13]. We are expecting a higher value of shallow trap level energy in our multilayered films compared to the pure BST films since the leakage current density is reduced in BST/ZrO$_2$ multilayered films due to the presence of interfaces. It is observed that the trap level energies for both Poole-Frenkel and SCLC mechanisms increase with the increase of individual ZrO$_2$ layer thickness. However, a gradual change in trap level energy with ZrO$_2$ layer thickness is not observed when comparison takes place between BZ1 (sub layer ZrO$_2$ thickness = 110 nm) and BZ4 (sub layer thickness of ZrO$_2$ = 27.5 nm) films. In order to fully understand the dependence of trap level energies on individual ZrO$_2$ layer thickness, further studies on BST/ZrO$_2$ multilayer films having different sub layer ZrO$_2$ thickness in the range between 27.5 and 110 nm are required.

From figure 2(b), it is also observed that the slope is close to 1 in the low electric field region. Therefore, it is possible that some other mechanism is contributing to the leakage current in the low field region rather than PF or SCLC mechanism. It is also known that Ohmic conduction process is observed when the volume generated intrinsic free carrier density, n_0, is higher than the injected free carrier density, n_i. It is observed that the slope of the ln J vs. ln E curve is slightly smaller than 1 for the multilayer films, in the low field region, which probably resulted from the ferroelectric polarization of BST thin film [14].

As the leakage current density is temperature dependent (Fig.1(a)) also in the low field region, the direct tunneling mechanism is ruled out. At this low field regime, the possible leakage mechanism is Ohmic conduction which is carried by thermally excited electrons hopping from one state to the next. This Ohmic current is given by [10]

$$J \sim E \exp(-E_a / kT) \quad (6)$$

where E is the electric field, T is the device temperature, and E_a is the thermal activation energy of conduction electrons.

To further confirm the dominance of Ohmic conduction mechanism, the I-V data is plotted as ln J versus ln E according to Eq. (6) and is shown in Fig. 2(b). It can be seen that the I-V data in the low field regime perfectly fits to a straight line with a slope ~ 1. Hence, Ohmic conduction is the dominant mechanism in the low electric field region. Therefore, one can conclude that the conduction process through BST/ZrO$_2$ multilayered films is Ohmic in the low field region and

SCLC mechanism in the medium field region whereas Poole-Frenkel is the dominant conduction mechanism in the high field region.

In the low field region where Ohmic conduction is dominant, the plot of ln J versus $1000/T$ gives a straight line at a fixed electric field as per Eq. (6) and the slope determines the thermal activation energy (E_a) for the Ohmic conduction mechanism. The activation energy was determined at a fixed electric field for both pure BST film and BST/ZrO$_2$ multilayered films. Figure 3(b) shows the plot for ln J versus $1000/T$ and the activation energy comes out to be as 0.05, 0.07 and 0.17 eV for B0, BZ1 and BZ4, respectively. The Ohmic current for BZ4 (inset in Figure (2b)) is lower compared to that of B0 and BZ1 films due to its higher activation energy for conduction electrons (0.17 eV).

From figure 2(b), the slope is higher than 2 in the high field region and the data points for this regime are already plotted as ln (J/E) vs. $E^{1/2}$ in Figure 1(b) and it fits to a straight line satisfying the Poole-Frenkel conduction mechanism in the high field region.

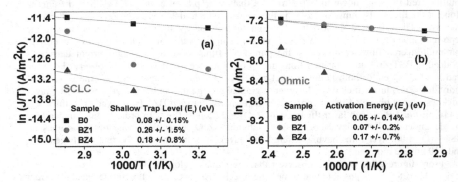

Figure 3. ln (J/T) vs. 1000/T (a) and lnJ vs. 1000/T (b) for B0, BZ1 and BZ4 heterostructured thin films.

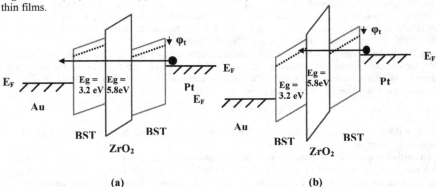

(a)   (b)

Figure 4. Energy band diagram of the Au/ BST/ZrO$_2$/BST/Pt capacitor at low field (a) and high field showing Poole-Frenkel conduction mechanism (b).

To physically explain the possible origin of the conduction mechanism in BST/ZrO$_2$/BST heterostructured thin films, we refer to the energy band diagrams in figure 4. Figure 4 (a) shows the band diagram at low field region. When low positive field is applied to the top electrode (Au), band bending occurs in the dielectrics [15]. The injected electrons from the bottom electrode (Pt) passes through all the three layers of heterostructured films and do not enter the trap level below the conduction band as shown in figure 4 (a). Therefore, Ohmic conduction is the dominant mechanism in the low field region [16]. When medium field is applied to the top electrode, more band bending occurs (not shown here). Thus, shallow electron trap level below the conduction band will participate in the conduction process contributing to the SCLC mechanism. However, when high positive field is applied to the Au electrode, some more band bending occurs in the dielectric layers compared to medium field conduction. Therefore, the injected electrons from the Pt electrode will enter the deep electron trap level and get trapped followed by detrapping to the conduction band in the dielectric (figure 4 (b)) and hence contributing to the Poole-Frenkel leakage current [16,17]. Thus, we will conclude that the electron trapping and detrapping occur from deep trap levels for the high positive field and therefore Poole-Frenkel mechanism is dominant in this region.

CONCLUSIONS

In summary, the carrier transport mechanisms are studied for BST films and BST/ZrO$_2$ heterostructured thin films fabricated on Pt/Ti/SiO$_2$/Si substrates by a sol-gel process. It is observed that in these films, Poole-Frenkel mechanism is dominant in the high field region with deep level traps whereas SCLC mechanism contributes in medium field region with shallow trap levels. Ohmic conduction process is dominant in the low field regime. All the deep trap, shallow trap level energies and activation energies are changing with the sub layer ZrO$_2$ thickness.

ACKNOWLEDGEMENTS

The authors would like to thank for the support from National Renewable Energy Laboratory, Colorado, USA and New Jersey Institute of Technology, New Jersey, USA.

REFERENCES

1. L. Fang, M. Shen, J. Yang, and Z. Li, solid state commun. **137**, 381 (2006).
2. C. Wenger, M. Albert, B. Adolphi, et al., Materials Science in Semiconductor Processing **5**, 233 (2003).
3. M. Ohishi, M. Shiraishi, K. Ochi, Y. Kubozono, and H. Kataura, Appl. Phys Lett. **89**, 203505 (2006).
4. Y. -B. Lin and J. Ya-min Lee, J. Appl. Phys. **87**, 1841 (2000).
5. Neil H. E. Weste and D. Harris, CMOS VLSI Design, A circuit and system perspective, 3rd Edition, (Addison-Wesley, Boston, 2005).
6. V. Reymond, D. Michau, S Payan and M Maglione, J. Phys. Condens. Matter **16**, 9155 (2004).
7. S. K. Sahoo, D. C. Agrawal, Y. N. Mohapatra, S. B. Majumdar, R. S. Katiyar, Appl. Phys. Lett. **85**, 5001 (2004).
8. M. Jain, S. B. Majumdar, R. S. Katiyar, and A. S. Bhalla, Thin Solid Films **447**, 537 (2004).
9. S. M. Sze, *Physics of Semiconductor Devices*, 2nd ed., Wiley-Interscience, (1981).

10. S. K. Sahoo, D. Misra, D. C. Agrawal, Y. N. Mohapatra, S. B. Majumder, and R. S. Katiyar, J. Appl. Phys. **108**, 074112 (2010).
11. M. Jain, S. B. Majumder, Yu. I. Yuzyuk, R. S. Katiyar, A.S. Bhalla, F.A. Miranda, F.W. Van Keuls, Ferro. Lett. Sectn. **30**, 99 (2003).
12. S. K. Sahoo, R. P. Patel, and C. A. Wolden, Appl. Phys. Lett. **101**, 142903 (2012).
13. C. –J. Peng and S. B. Krupanidhi, IEEE, 460 (1995).
14. S. Y. Wang, B. L. CHENG, C. Wang, S.Y. Dai, H. B. Lu, Y. L. Zhou, Z. H. Chen, G. Z. Yang, Appl. Phys. A **81**, 1265 (2005).
15. N. Alimardani, E. W. Cowell, J. F. Wager, J. F. Conley, D. R. Evans, M. Chin, S. J. Kilpatrick, and M. Dubey, J. Vac. Sci. Technol. **A 30(1)**, 01A113-1 (2012).
16. S. K. Sahoo and D. Misra, J. Appl. Phys. **110**, 084104, (2011).
17. S. K. Sahoo and D. Misra, Appl. Phys. Lett. **100**, 232903 (2012).

Mater. Res. Soc. Symp. Proc. Vol. 1547 © 2013 Materials Research Society
DOI: 10.1557/opl.2013.607

Fabrication of $0.6(Bi_{0.85}La_{0.15})FeO_3$-$0.4PbTiO_3$ Multiferroic Ceramics by Tape Casting Method

Guoxi Jin, Jianguo Chen, Shundong Bu, Dalei Wang, Rui Dai and Jinrong Cheng[*]
School of Materials Science and Engineering, Shanghai University, 200072, P.R. China
[*]corresponding author: jrcheng@shu.edu.cn

ABSTRACT

The $0.6(Bi_{0.85}La_{0.15})FeO_3$-$0.4PbTiO_3$ (BLF-PT) ceramics were prepared by tape casting method. Effects of binder (polyvinylbutyl dibutyl PVB), plasticizer (phthalate-polyethylene glycol DBP-PEG) and dispersant (triethylolamine, TEA) concentration on the rheological properties of BLF-PT slurry were investigated. The optimized component ratio for ceramics powders, binder, plasticizer, dispersant and solvent (ethanol, EtOH) in the slurry was 50 wt.%, 4 wt.%, 6 wt.%, 1 wt.% and 39 wt.%. The dielectric constant ε_r, loss $tan\delta$, and remnant polarization P_r of BLF-PT ceramics laminated from the tapes were 525 (1 kHz), 1.7% (1 kHz) and 30 $\mu C/cm^2$ (45 kV/cm), respectively, which were comparable to those of BLF-PT ceramics prepared by traditional solid state reaction method.

INTRODUCTION

Tape casting method is capable of fabricating flexible and low-cost thick films with widest thickness range [1]. Compared with conventional process, tape casting owns the great advantage in forming large-area, thin and flat ceramic parts. And it has been widely used in electronics industry for capacitors, piezoelectric and electrostrictive devices, ferromagnetic memories, ceramic substrates, fuel cells and so on. The tape casting process involves casting the slurry onto a flat carrier surface [2]. The slurry usually consists of ceramic powders with appropriate additions of solvents, plasticizers, dispersants and binders. The solvents evaporate to leave a relatively dense flexible sheet or ceramic tape that may be stored on rolls or stripped from the carrier in a continuous process [3].

Recently, materials based on the $BiFeO_3$-$PbTiO_3$ (BF-PT) system have drawn considerable attentions because of their high Curie temperature and the coexistence of ferroelectric and ferromagnetic orderings in single phase at room temperature, exhibiting potential applications in spintronics, information storage, sensing and actuation devices [4-7]. However, low resistivity and weak ferromagnetic property at room temperature makes it difficult to be used in practical applications [8-10]. Our previous work showed that La modified BF-PT ceramics exhibited enhanced multiferroic and highly insulated properties simultaneously at room temperature [11].

In this paper, La modified BF-PT ceramics were synthetized by tape casting method. Effects of binder, plasticizer, and dispersant concentrations on the rheological properties of BLF-PT slurry were discussed, and then electrical properties of ceramics derived form the tapes were characterized.

EXPERIMENT

Analytical-grade powders of Bi_2O_3, TiO_2, La_2O_3, Fe_2O_3 and PbO were used as starting materials to synthetize $0.6(Bi_{0.85}La_{0.15})FeO_3-0.4PbTiO_3$ solid solution. Firstly, they were blended and adequately ball-milled with deionized water for 24 h. Then, the mixture was dried and calcined at 750 °C for 4 h in a sealed alumina crucible. The processes of ball-milling and calcination were repeated twice to obtain the single phase powders. The synthetized powders were sieved through 140-mesh screen to obtain a narrow size distribution.

The prepared powders were mixed in ethanol with TEA (dispersant) for 2 h first. Then PVB (binder) and PEG-DBP (plasticizer) were added in the slurry, and followed by mixing for another 12 h to make the slurry fine and uniform. Different concentrations of binder, plasticizer and dispersant were introduced in the slurry, and then the rheological properties of these slurries were measured by a viscometer (Thermo Viscotester 550, USA). The slurry with proper concentrations of ceramic powders, solvents, plasticizers, dispersant and binders was casted on a plate glass surface by a blade to form tape with thickness of 400 μm after ball-milling. Tape drying was carried out at room temperature in the open air without blowing. The dried green sheets were punched, laminated and cold isostatic pressed at 200 MPa for 2 min to obtain homogeneous disks of 12 mm in diameter. The pressed tablets were heated in air at 600 °C for 3 h to remove all organic additives and then sintered at 1040 °C in air for 0.8 h. Finally, the ceramics were polished and coated with sliver electrodes on both sides. Dielectric and ferroelectric properties were measured by a precision impedance analyzer (Agilent HP4294A, USA) and a ferroelectric measurement system (Radiant Premier II, USA), respectively.

RESULTS AND DISCUSSION

The quality of the slurry affects the toughness and flatness of the tapes significantly. The expected slurry shows the rheological property that its viscosity is not very sensitive to the variation of shear rate. Viscosity of such slurry can be observably decreased with a blade passing and rapidly recovering the original viscosity to keep the shape. Therefore, different concentrations of binder, dispersant and plasticizer were added into the slurry to adjust the slurry rheological property till an optimal value.

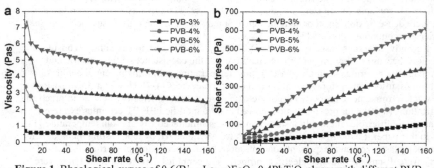

Figure 1. Rheological curves of $0.6(Bi_{0.85}La_{0.15})FeO_3-0.4PbTiO_3$ slurry with different PVB concentrations

To investigate the effect of binder content on the stability of casting slurry, the rheological behavior of slurries with a solid loading of 50 wt.% and 44~38 wt.% EtOH were measured as a function of shear rate and shown in Figure 1. The total concentrations of PVB/DBP-PEG were from 6 to 12 wt.%, maintaining the binder/plasticizer ratio as 1:1. As shown in Figure 1, some viscosity curves show a smooth and steady change with increasing shear rate. However, slurries with over 5 wt.% PVB show an obvious sliping tendency as the increasing shear rate which indicates unhomogenization in slurry. That may be caused by undissolved binders enhancing the internal friction. As a result, excess binder agglomerates were formed in high PVB content slurries. Compared with the PVB concentration of 3 wt.%, the slurry with 4 wt.% PVB shows a higher static viscosity which indicates a better film shape and flatness keeping after tape casting. For these reasons above, a concentration of 4 wt.% PVB was selected as the optimum content for slurry preparation.

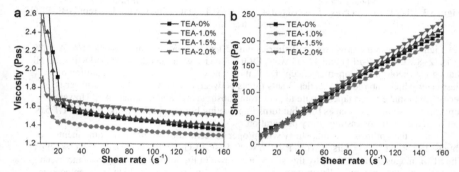

Figure 2. Rheological curves of $0.6(Bi_{0.85}La_{0.15})FeO_3-0.4PbTiO_3$ slurry with different TEA concentrations

Further improvement was carried out by optimizing the dispersant content up to complete stabilization. Figure 2 shows the flow curves with different concentration of TEA from 0 to 2 wt.%. Slurries were mixed with 50 wt.% powders, 4 wt.% PVB, 4 wt.% DBP-PEG and 42~40 wt.% EtOH. All the viscosity curves show the same tendency as binder adjustment. When the content of TEA was 1 wt.% and other components did not change, the slurry displayed the lowest viscosity. It is interesting that slurry viscosity increased with the excessive content of TEA, slurries still kept stable shear-thinning behavior. As a result, a concentration of 1 wt.% TEA was selected to continue.

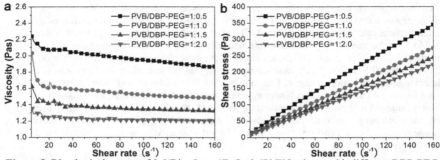

Figure 3. Rheological curves of $0.6(Bi_{0.85}La_{0.15})FeO_3-0.4PbTiO_3$ slurry with different DBP-PEG concentrations

Figure 3 shows the flow curves of slurries with 50 wt.% solid loading, 1 wt.% TEA, 4 wt.% PVB, 2~8 wt.% DBP-PEG and 42~37 wt.% EtOH. Slurries with the ratio of PVB/DBP-PEG all share great rheological properties except 1:0.5, it may be caused by insufficient addition of plasticizer which can reduce Van der Waals forces between binder molecules. Plasticizer is necessary to make green tapes flexile to peel from substrate and free laminate [12]. However superfluous use improves tapes little but turns into heating burden. In this work, ratio of 1:1.5 is considered enough for casting slurry preparation.

Hereto, the optimized casting slurry with proper concentration of different components was obtained as 50 wt.% powders, 4 wt.% PVB, 6 wt.% DBP-PEG, 1 wt.% TEA and 39 wt.% EtOH. The SEM images of tape casted by this slurry are shown in Figure 4. SEM images of the green tapes indicate the uniformly distribution of solid particles in dried tapes which is advantageous for lamination in a continuous process.

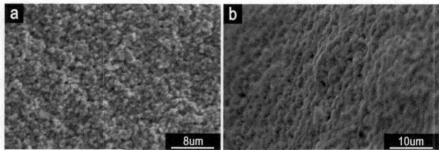

Figure 4. SEM images of $0.6(Bi_{0.85}La_{0.15})FeO_3-0.4PbTiO_3$ green tape casted by optimized slurry
a) top surface and b) fracture surface

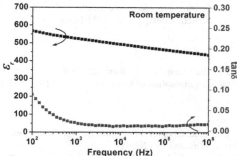

Figure 5. Dielectric constant and loss of $0.6(Bi_{0.85}La_{0.15})FeO_3$-$0.4PbTiO_3$ ceramic by tape casting as a function of frequency at room temperature

Figure 5 shows the dielectric constant ε_r and loss $tan\delta$ of BLF-PT ceramics prepared by casting tapes as a function of frequency from 10^2 to 10^6 Hz. The dielectric constant of the specimen is 525 at 1 kHz while the loss is 1.7%.

The ferroelectric hysteresis loop of BLF-PT is presented in Figure. 6. The remanent polarization P_r and coercive field E_c of 30 $\mu C/cm^2$ and 27 kV/cm. Ceramics prepared by traditional solid state reaction share the dielectric and ferroelectric properties of ε_r= 500 (1 kHz), $tan\delta$= 3.2% (1 kHz), P_r= 35 $\mu C/cm^2$ and E_c= 25 kV/cm, respectively.

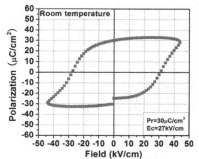

Figure 6. P-E loop of $0.6(Bi_{0.85}La_{0.15})FeO_3$-$0.4PbTiO_3$ ceramic prepared by tape casting

CONCLUSIONS

The casting slurry composition for $0.6(Bi_{0.85}La_{0.15})FeO_3$-$0.4PbTiO_3$ was detailedly discussed in this letter. By analyzing the rheological properties and tapes quality of different component concentration, dosage of each kind of additives was selected. The best composition of $0.6(Bi_{0.85}La_{0.15})FeO_3$-$0.4PbTiO_3$ casting slurry is 50 wt.% solid content, 4 wt.% PVB, 6 wt.% DBP-PEG, 1 wt.% TEA and 39 wt.% EtOH. Ceramics laminated by such tapes exhibit comparable electrical performances to those prepared by conventional method. In addition, tape casting shows more convenient forming process than conventional craft, it will be essential to the further research of multilayered structure design in BLF-PT based ceramics and devices.

ACKNOWLEDGMENTS

This work was financed by the Shanghai education development foundation under grant No. 08SG41, National Nature Science Foundation of China under grant No. 50872080.

REFERENCES

[1] R. E. Mistler and E. R. Twiname, Chapter 4 in *Tape Casting: Theory and Practice*, The American Ceramic Society, Westerville, (2000).
[2] K. Singh, N. S. Negi, R. K. Kotnala, M. Singh, *Solid State Commun.*, **148**, 18-21 (2008).
[3] T. Leist, K. G. Webber, W. Jo, E. Aulbach, *ACTA Mater.*, **58**, 5962-5971 (2010).
[4] W. Eerenstein, N. D. Mathur, J. F. Scott, *Nature*, **442**, 759-765 (2006).
[5] M. Fiebig, *J. Phys. D.*, **38**, R123-52 (2005).
[6] M. Li, M. Ning, Y. Ma, Q. Wu, C. K. Ong, *J. Phys. D.*, **40**, 1603-1607 (2007).
[7] J. R. Cheng, R. Eitel and L. E. Cross, *J. Am. Ceram. Soc.*, **86** [12], 2111–2115 (2003).
[8] D. I. Woodward, I. M. Reaney, R. E. Eitel, C.A. Randal, *J. Appl. Phys.*, **94**, 3313 (2003).
[9] T. P. Comyn, S. P. McBride, A. J. Bell, *Mater. Lett.*, **58**, 3844 (2004).
[10] A. Navarro, J. R. Alcock, R.W. Whatmore, *J. Eur. Ceram. Soc.*, **24**, 1073-1076 (2004).
[11] G. Y. Shi, J. G. Chen, J. R. Cheng, *Curr. Appl. Phys.*, **11**[3], 251-254 (2011)
[12] J. K. Song, W. S. Um, H. S. Lee, M. S. Kang, K. W. Chung, J. H. Park, *J. Eur. Ceram. Soc.*, **20**, 685-688 (2004).

Materials for Energy and Electronic Devices

Mater. Res. Soc. Symp. Proc. Vol. 1547 © 2013 Materials Research Society
DOI: 10.1557/opl.2013.636

Solution Processed TiO2 Nanotubular Core with Polypyrrole Conducting Polymer Shell Structures for Supercapacitor Energy Storage Devices

Navjot K. Sidhu[1,2], Ratheesh R. Thankalekshmi[1,2], A.C.Rastogi[1,2]

[1]Electrical and Computer Engineering Department,
[2]Center for Autonomous Solar Power (CASP),
Binghamton University, State University of New York, Binghamton, NY13902, U.S.A.

ABSTRACT

Ordered one dimensional polypyrrole conducting polymer structure as a shell over TiO2 nanotube arrays at the core were formed by pulsed current electropolymerization. TiO2 nanotubes with rippled wall structure are designed by action of water in the anodizing medium. This provides open tube structure supporting short diffusion length and increased accessibility of ions involved in redox transition for energy storage. Electrochemical properties evaluated by cyclic voltammetry and electrochemical impedance spectroscopy show specific capacitance of 34-44 mF.cm^{-2} and extremely low bulk and charge transfer resistances.

INTRODUCTION

Ordered one dimensional TiO2 nanostructures show potential for use in energy devices such as in dye-sensitized solar cells [1], Li-ion intercalation battery anode [2] and supercapacitors [3]. Polyaniline (PANI) and polypyrrole (PPy) conducting polymers having reversible redox properties hold special interest as low-cost solution-processed electrodes for supercapacitors but mostly have been studied in the macroporous two-dimensional (2D) structures. Recently, low dimensional conducting polymers such as nanofibres and nanowires etc are being studied aimed at improved and stable supercapacitor performance [4]. These conducting polymer structural forms created by interface polymerization or using micelle soft templates are in randomly dispersed forms. Ordered one-dimensional conducting polymer nanostructure appear more promising since these minimize ion diffusion distances and maximize ion accessible surface area for efficient redox processes. In this work, a novel structure having polypyyrole conducting polymer shell structure over ordered TiO2 nanotube arrays at the core has been created by pulsed current electropolymerization in aqueous media. This paper reports synthesis, structure and electrochemical properties relevant to the supercapacitor energy storage.

EXPERIMENT

Fabrication of TiO2 Core and PPy- shell structure

One dimensional structure of vertical nanotubes of TiO2 was fabricated by anodization of a Ti foil in 0.5 wt% NH4F with ethylene glycol as a polar organic electrolyte. Anodization was performed in an electrochemical cell with Pt sheet cathode using a constant potential of 20 V. To fabricate the supercapacitor device, we created ordered nanotubes arrays with rippled wall structure by controlled water content of 20% and 30 % *vol* in the anodization medium. This serves as a core for fabricating electrode for supercapacitor device. For creating PPy shell over core of TiO_2 nanotube array, it is important to synthesize the doped- polypyrrole film over the rippled TiO_2 nanotubes. This was achieved by *insitu* pulsed current electropolymerization of pyrrole monomer in an aqueous medium [5]. Electro-polymerization was carried out in a 0.1M H_2SO_4 medium in the presence of surfactant 0.01M SDS (sodium dodecyl sulfate) and 0.2M pyrrole (Py) monomer in an electrochemical cell. Unipolar square current pulses designed with an on-duration of 10 ms, inter-spaced by a 100ms pulse off time to realize site-selective PPy film formation. Typically, the pulsed current density of $4mAcm^{-2}$ based on the geometric area of TiO_2 nanotubes sheets was used. Polymerization of pyrrole monomer takes place directly over the TiO_2 nanotubes array and to build a reasonable PPy shell thickness; these current pulses were sequentially applied for over 30k cycles.

Electrochemical characterizations

Pseudocapacitive properties of Ti sheet backed TiO_2-PPy core-shell electrodes were studied in an electrochemical cell by cyclic voltammetry (CV) using electrochemical analyzer (Solartron 1287A) in a 0.5M H_2SO_4 solution having a Pt counter and Ag/AgCl reference electrode. The impedance spectra were measured in the frequency range 10mHz-100kHz at a 10mV signal level using gain-phase analyzer (Solartron 1260A). These data were analyzed using conventional methods to elucidate the electrochemical performance of the TiO_2-PPy core-shell electrodes.

DISCUSSION

TiO_2 nanotube core structure

Figure 1 shows scanning electron micrograph of the vertical TiO_2 nano-tube arrays with average pore diameter ~ 75 nm formed with 20% and 30% water addition. The TiO_2 nanotube structure was created uniformly without any grass- like structures on top surface. High water content based TiO_2 nanotube core resulted in controlled ripple like multi-wall geometry with more pore-wall openings. Such a structure forms by additional oxygen ions which increase the dissolution of TiO_2 layer back into electrolyte resulting in the ripple formation on top surface tubes as shown in the figure 1(b). The tube wall geometry is an important factor in determining the electrical properties of such arrays forming the core of the supercapacitor device fabricate by imparting more volumetric access to ions across PPy shell.

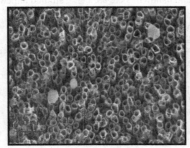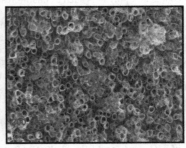

Figure 1. SEM images of top surfaces of as-anodized TiO_2 nanotube array prepared by anodic oxidation with addition of water content (a) 20 % *vol* (b) 30 % *vol*

Growth of TiO₂/PPy Core - Shell Structure

Figure 2 shows the SEM images of the PPy layer forming the shell structure over TiO_2 nanotubes. In 20% water formed TiO_2 nanotubes, having smaller inter-tube gaps, the PPy shell over individual tubes coalesce and forms a continuous electrically connected network shown in figure 2(a). No or little electrodeposition of PPy polymer film occurs in inner walls of these tubes. PPy deposited over TiO_2 nanotube core created with 30% water shows open pore shell structure with rippled walls of nanotubes are selectively coated while the inter-tube gaps in the PPy shells are not fully covered as shown in the figure 2(b). Anonic surfactant (SDS) helps forming a conformal layer of PPy over TiO_2 array in creation of such a core-shell structure. A sufficiently small SDS concentration lower than the critical value for micelles formation has been used in the process. Thus, monomeric amiphiphile surfactant having higher ionic mobility than that of the micelles combines either with the Py monomer by establishing ionic bond with polycation of Py or gets adsorbed on PPy surface due to its hydrophobic property [6]. This hinders the solubility of PPy-SDS dimer and the electrostatic force between the TiO_2 nanotube walls. Consequently, cationic polymers (molecular anchor) lead to the preferential deposition of PPy along the outer walls and thus in the gap between them rather than the inner walls of nanotubes. Diffusion rate of PPy inside the nanotubes is very low as compared to clustering of PPy around the corners of tubes. Pulse polymerization promotes a slow deposition of polymer epitaxially between the walls which is carried out for over an hour. Wetting properties of inner pores of TiO_2 nanotubes are also different from those at the outer walls and the open space between the PPy shells of adjacent tubes may result in open network structure. Highly ordered TiO_2 structure coated with PPy provides shorter ion diffusion length and fast electron transportation in contrast to the conducting polymer matrix deposited on flat 2-D substrates.

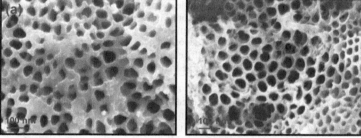

Figure 2. SEM images of top surfaces of as-anodized TNT covered by conducting polypyrrole (PPy) coating by pulse polymerization. (a) 20 % Core Shell, (b) 30 % Core Shell

Single Electrode Impedance Spectroscopy

Impedance Spectroscopy is a powerful tool to study the electrical properties of electrodes, electrolyte and their interfaces. Figure 3 compares the impedance spectra of 20% and 30% TiO_2 nanotube core-PPy shell structured electrodes. In both cases, a sharp increase in the imaginary impedance (Z'') in the lower frequency range testifies to a highly capacitive nature of the TiO_2 nanotube-PPy core-shell electrodes. Impedance spectra from the TiO_2 nanotube array core without PPy shell show dispersive impedance and no capacitive property. Using the relation, $C =- 1/2\pi f.Z''$, specific capacitance of 20% TiO_2 nanotube core-PPy shell electrode was calculated

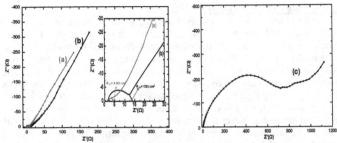

Figure 3. Nyquist Plot of TiO₂-PPy core-shell electrode in the 10 mH-100 kHz frequency range (a) 30 % Core-Shell, (b) 20 % Core-Shell (c) 20% Core

as 34 mFcm⁻². The 30% TiO₂ nanotube core-PPy shell electrode shows a higher 44 mFcm⁻² specific capacitance. Semicircle features in impedance spectra at high frequency are due to electrode-electrolyte interface resistance effects shown in inset of figure 3(a). In the 20% TiO₂-PPy core- shell, existence of high frequency region semicircle showed bulk resistance R_b=1.5Ω cm² and charge transfer resistance R_{ct}=13 Ω cm², whereas in 30% TiO₂ nanotube-PPy core- shell, bulk resistance is almost the same but R_{ct} value decreases by almost three times. This is due to fast Faradaic redox reactions over large area open surface network of porous electrode in the 30% TiO₂ naotube core-PPy shell electrode structure as shown by the microstructure study. Nyquist plot shown in inset reflects the Warburg /diffusion impedance which is higher for 20 % than for 30 % TiO₂ nanotube core-PPy shell electrode. Dopant ions encounter more resistance in diffusing towards the pores in 20% TiO₂ nanotube core-PPy shell electrode as this network is more compact. The TiO₂ nanotube core assembly significantly boosts the performance of supercapacitor electrode. The PPy film similarly deposited over thin graphite sheets shows R_b= 10 Ω cm² and R_{ct}= 15 Ω cm². Compared to 2-D PPy eledotes, bulk resistance of TiO₂ nanotube-PPy core-shell electrodes decreases by 10 and charge transfer resistance by 4 times.

Cyclic Voltammetry

Cyclic voltammetry (CV) measurements were used to study electrode kinetics in a three electrode system. Voltammetry is carried out between -0.25 V to 0.5 V at different scan rates 10, 20, 50 and 100mVs⁻¹ and the results are depicted in figure 4. Voltage (V_1= -0.25) is selected due to negligible current in order to avoid oxidation and reduction at initial stage and voltage (V_2=0.5) is selected such that this range covers the interested oxidation and reduction process for working electrode without any chemical reaction. Anodic current (positive value) and cathodic current (negative value) represents oxidation and reduction respectively at the working electrode. Based on plot in figure 4(a) specific capacitance for both electrodes was evaluated by the equation $(I_a+I_c)/2s$, where I_a and I_c represent anodic and cathodic currents, respectively and $s = (dv/dt)$ is the voltage sweep rate. Specific capacitance of 33mF cm⁻² for 20 % and 42mF cm⁻² for the 30% TiO₂ nanotube core-PPy shell electrodes is obtained consistent with the impedance data. TiO₂ nanotube interior contributes little specific capacitance which is attributed to electric double layer and most contribution originates from the Faradaic redox reactions in the PPy shell. CV plots of 20% and 30% TiO₂ nanotube-core PPy shell electrodes differ substantially as seen in figure 5. The 30% core-shell electrode yield higher unit area current

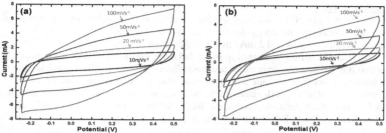

Figure 4. Cyclic voltammetry (CV) curves of TiO$_2$ nanotube-PPy core-shell electrodes measured at different scan rates (a) 30% core-shell (b) 20% core-shell

and more rectangular and symmetrical voltammogram. This results from better electrical conduction and anion diffusion deeper over nanotube length and across wider surface regions through the rippled core walls in the doping-dedoping process. The CV plots of 20% core-shell electrode are not as nearly rectangular indicating a higher resistance due to constrained ionic diffusion though TiO$_2$ nanotube due to coalesced shell structure at the outer top regions of nanotubes. In this case tilted voltammograms arise due to delayed current response from top and the regions across the length of nanotubes. Referring to figure 4, the current increases and the specific capacitance decreases with increase in the voltage scan rate. This results from dependence of anion insertion and extraction on the diffusion to the active sites on the PPy shell. The diffusion process is quantified by Randles-Sevcik analysis of peak current values at different scan rates. Using equation, $I_p = 2.69 \times 10^5 \cdot n^{3/2} A D^{1/2} v^{1/2} C$, where I_p is peak current, n electrons involved in redox process, A electrode area, D Diffusion coefficient, v scan rate, and C is ionic concentration, ionic diffusion behavior of ions in both the core-shell structures is determined. Accordingly, 30 % TiO$_2$ nanotube core-pPy shell electrode shows faster diffusion kinetics in agreement with the morphology studies related to an open pore core-shell structure. Increase of peak current implied improved Faradic reactions and enhanced charge transportation resulted in capacitive behavior. Highly ordered TiO$_2$ nanotube cores with PPy shell allows direct movement of electrons around the walls and ions through the pores of the tubes.

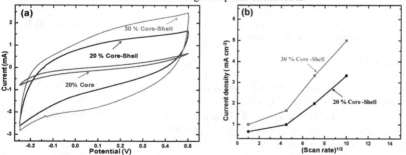

Figure 5. (a) CV curves of 20% core, 20% and 30% TiO$_2$ nanotube-PPy core-shell electrodes at scan rate 20mV s^{-1} (b) Randles-Sevcik plots for both the electrodes

Charge -Discharge Tests

Charge storage capability, life expectancy and electrical parameters such as power, energy of the electrodes are evaluated from charge-discharge (CD) tests. This study was carried out at different current densities, 0.5, 1 and 2 mA.cm^{-2}. Voltage range is selected between 5mV to 0.5V. ESR is calculated from the voltage drop in the discharge curve. CD curves are almost symmetrical and linear. Capacitance is calculated from discharge curve excluding voltage drop by standard analytical methods. Voltage drop across the discharge curve tells about pores resistance. From this analysis, specific capacitance of 38 mF cm^{-2} and 43 mFcm^{-2} are determined for 20% and 30% core-shell electrodes. Galvanic charge discharge test shows 30% core shell performs better than 20% core shell consistent with impedance and CV test results.

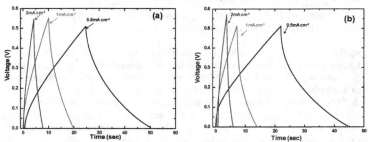

Figure 6. CD curves at different current density (a) 30% core-shell (b) 20% core-shell

CONCLUSIONS

Ordered 1-D polypyrrole conducting polymer structure formed as shell over TiO$_2$ nanotube arrays are investigated as supercapacitor electrode. With rippled wall core, high specific capacitance of 44 mF.cm^{-2} and low~4 Ω.cm^2 charge transfer resistance are obtained which show such nanostructures have considerable potential for supercapacitor energy storage devices.

ACKNOWLEDGMENT

This work was supported by the Office of Naval Research (ONR) under contract N00014-11-1-0658 which is gratefully acknowledged.

REFERENCES

1. J.R. Jennings, A.Ghicov, L.M. Peter, P.Schmuki, et.al. *J. Am. Chem. Soc.,* **130**, 13364 (2008).
2. M. G. Choi, Y.-G Lee, S.-W Song, K.M Kim, *Electrochim Acta,* **55**, 5975 (2010).
3. X. Lu, G. Wang, T. Zhai, M. Yu, J. Gan, Y. Tong and Y. Li, *Nano Lett.,* **12**, 1690 (2012).
4. S Sahoo, S. Dhibar, C.K. Das, *eXpress Polym. Lett.,* **6**, 965 (2012).
5. R. K. Sharma, A. C. Rastogi, and S. B. Desu, *Electrochem. Comm.,* **10**, 268 (2008).
6. M. Omastovaa, M. Trchova, J. Kovarova, J. Stejskal, *J. Synth. Met.* **138**, 447 (2003).

Mater. Res. Soc. Symp. Proc. Vol. 1547 © 2013 Materials Research Society
DOI: 10.1557/opl.2013.679

Dual Function Polyvinyl Alcohol Based Oxide Precursors for Nanoimprinting and Electron Beam Lithography

J. Malowney[1,2], N. Mestres[1], X. Borrise[2,3], A. Calleja[1], R. Guzman[1], J. Llobet[2], J. Arbiol[1,4], T. Puig[1], X. Obradors[1] and J. Bausells[2]

[1] Institut de Ciència de Materials de Barcelona (ICMAB-CSIC), Bellaterra, 08193, Spain
[2] Institut de Microelectrònica de Barcelona (IMB-CSIC), Bellaterra, 08193, Spain
[3] Catalan Nanotechnology Institute (ICN), Bellaterra, 08193, Spain
[4] Catalan Institution for Research and Advanced Studies (ICREA), Barcelona, 08010, Spain

ABSTRACT

Ordered arrays of crystalline complex oxides nanostructures were synthesized onto single crystal insulating substrates using aqueous polyvinyl alcohol based electron beam resist precursors. The irradiated zones are insoluble in water (negative-tone resist) due to the electron induced cross linking of polyvinyl alcohol. The subsequent high temperature treatment of the developed precursor samples leads to the formation of ordered arrays of nanodots for low irradiation doses. For high irradiation dosages, epitaxially and oriented nanowires are obtained. These same precursors were shown to be nanoimprintable on single crystal substrates. This allows for future dual processing of a single precursor film gaining nano-structuration from both electron beam and nanoimprint lithography methods.

INTRODUCTION

In the last years, investigations concerning functional oxide nanostructures have received a great deal of attention for the development of future technologies like magnetic memories or new types of sensors and logic devices [1,2], because nanoscale complex oxides can display different and improved magnetic, electrical, optical and mechanical properties compared to their bulk counterparts. Since many functional properties rely on epitaxial order and crystal orientation (e.g., piezoelectric and high-T_C superconducting properties), growth methods that are able to provide ordered arrays of epitaxial nanostructures are highly interesting. Several recent studies in our group have grown epitaxial single crystalline functional oxide nanostructures on single crystal substrates [3-5]. However, these nanostructures were generated with no precise position control. Through the use of lithography, nanostructures may be written in arrays and other configurations. $La_{0.7}Sr_{0.3}MnO_3$ (LSMO) was the functional complex oxide we focused in this work due to its interesting physical properties. It has a Curie temperature of 360 K, the ability to change electrical resistance dramatically under applied magnetic field, half metallicity for spin conduction, a highly anisotropic magnetic nature, and the ability to alter its physical properties when strained [6,7].

Chemical solution growth methods based on an aqueous solution laden with soluble metal salts is thought to be an inexpensive and easily scalable growth approach. Our aim was to combine a polyvinyl alcohol (PVOH) based electron sensitive precursor resist with electron beam lithography to obtain epitaxial LSMO patterns with sub-micrometer-scale precision offering many attractive opportunities in the fabrication of metal oxide based devices. PVOH based resists has been reported to possess both positive and negative imaging properties depending on the e-beam irradiation doses [8]. PVOH crosslinking leads to a difference in solubility with respect to unexposed areas in the developer (distilled water); i.e., the irradiated and crosslinked PVOH areas are insoluble in water. The subsequent high temperature treatment would lead to phase formation and crystallization. This procedure was previously outlined by Wu et al., [8,9] to grow LSMO nanostructures. However the substrates used in that work were Si single crystals and hence the resulting patterned ceramics were typically amorphous or polycrystalline due to the large lattice mismatch existing between the oxide and the substrate. The other form of lithography explored in this work was nano-imprint-lithography (NIL) using the same kind of precursors as imprintable films.

EXPERIMENT

The electron beam resist was prepared by dissolving high purity (>99.99%) metal nitrate salts in Milli-Q water with a molar ratio of La:Sr:Mn=0.7:0.3:1. This solution was then combined with PVOH (molecular weight 18,000 g/mol, Sigma Aldrich) aqueous solution to assure a wt% of polymer between 2% and 15% in the final solution. The solution was then spun onto the insulating substrate. The substrates used were $SrTiO_3$ (STO), $LaAlO_3$ (LAO), and yttria stabilized ZrO_2 (YSZ). This generated a uniform homogeneous 200-300 nm thin film depending on the PVOH concentration over the insulating substrate ready for e-beam or NIL processing.

High-resolution e-beam lithography was performed by using the scanning electron microscopes of FEI Quanta 200F, Raith 150 Two Direct Write, and Zeiss LEO 1530. The standard electron beam conditions were a working distance of ten millimeters, an aperture of thirty micrometers, and a beam current of 150 pA. The substrate was then removed from the vacuum chamber and developed using pure water for 30 s. This development stage resulted in a cleaned substrate with hardened parts of the precursor film attached where electrons were focused. Finally, the sample was placed into a tube furnace in an oxygen rich atmosphere to anneal the hardened structures for LSMO phase formation and crystallization. The annealing temperatures were in the 900-1000 °C range. NIL experiments were performed in an Obducat commercial system able to process up to 4-inch wafers with nm feature sizes.

The intermediate and final nanostructures were observed with SEM as well as with atomic force microscopy (AFM). The AFM used was a Veeco Dimension 3100 by Digital Instruments.

DISCUSSION

We first demonstrated the capability to grow high quality epitaxial and ferromagnetic LSMO thin films on top of (001)-STO and -LAO substrates from our PVOH based resist.

Nanodots growth

After the LSMO precursor has been radiated with focused electrons at predetermined locations and durations, the non-radiated film must be removed. To do this, the sample was submerged in water which acts as a solvent. The resulting nanodots shape after water developing comes from the shape of the electron beam. At the very lowest dosages, the height grows slightly with an increase in dosage. However, it soon reaches an asymptote and thus a maximum height is reached at slightly higher dosages as can be seen on figure 1(a). This observation is explained by the premise that the more electrons focused onto a site the more that site will cross-link and adhere to the substrate. This occurs until a certain dosage is reached and thereafter the height of the nanoisland reaches a maximum due to the geometrical considerations of the film.

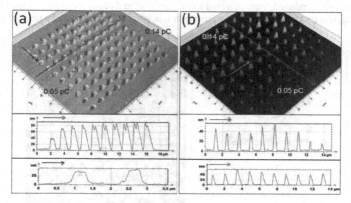

Figure 1. (a) Nanodots array after water development showing electron dosage dependence. The pitch was 1.25 μm, while the dosage ranged from 0.05 to 0.14 pC in increments of 0.01 pC. The substrate was STO. (b) The same sample after the final thermal treatment 4 h at 1000 °C.

The resulting morphology of the final nanoisland is a result of the amount of insoluble crosslinked precursor left on the substrate after water development. Figure 1(b) displays the resulting nanoislands after annealing the developed sample at 1000 °C, 4h. This gave between 15 nm and 40 nm high and ~ 150 nm wide oxide nanodots. The volume of the final nanodots is roughly 80% less than the volume of the non-annealed nanosites due to the combustion of the remaining organic matter and to compaction in the crystallization process.

Transmission electron microscopy (TEM) coupled with electron energy loss spectroscopy (EELS), a technique sensitive to atomic presence allows for the precise confirmation of the elemental composition inside the nanodot as well as determining the degree of epitaxy on the substrate. TEM measurements (not shown) revealed the growth of single crystal epitaxial oxide nanostructures. However, the results of EELS line scans performed along the nanodot showed that lanthanum and strontium were present in the nanodot however manganese was not. The absence of manganese in the nanodots indicates that the generated nanostructures were actually not LSMO, rather a lanthanum strontium oxide. This was not a wholly unexpected result due to other reported work showing the migration of manganese out of LSMO nanostructures [3]. The possibility that manganese is lost during the electron beam writing process, or the developing step should also be considered, and is under study.

Nanowires growth

While running high dosage and density tests for morphology of PVOH based nanodots on STO, it was discovered that at very high irradiation dosages (15 and 500 mC/cm^2) oriented nanowires formed after annealing. An example of this growth mode can be seen in figure 2, which consisted of a hundred nanodot array (10 x 10) spaced 750 nm apart and each single dot written with a dosage of 1500 pC on top a (001)-STO substrate. The nanowires grew at the edge of the area that was irradiated, were oriented along the [100] and [010] directions on the (001)-STO substrate, and had lengths up to 10 micrometers, heights of 25 ± 5nm and widths close to 200nm.

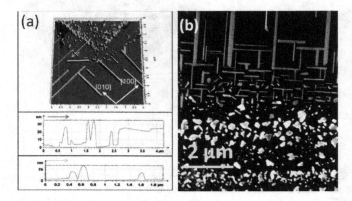

Figure 2. (a) AFM and (b) SEM images of nanowires generated by a high dosage electron irradiation on STO (001) after annealing at 1000 °C during 4 hours. Array size 10×10 with 1500 pC per site and 750 nm pitch.

In order to understand the transformation of these high dosage irradiated regions into nanowires more precisely, a detailed study by AFM imaging was made. We realized that the nanowires were actually formed in an area which was modified with enough secondary electrons to prevent substantial material loss when washed in developer. This behavior is probably enhanced by considering the charging ability of STO. In the AFM image in figure 2(a), we observe that at the top of the figure corresponding to the directly irradiated area, a random distribution of polycrystalline nanoparticles was formed with an average grain size around 60 nm. Surrounding this area, one can see where the majority of nanowires start to grow and that they do so along well defined directions. In the directly irradiated region with high dosage the polymer resist has been strongly affected during the irradiation process and some crystallization has been already induced [8]. As a consequence, the high temperature thermal treatment leads to the formation of a large number of random distributed nanocrystals in this area. In the secondary zone however, the cross linked polymer is retained after water development and there is enough undamaged precursor material to promote the growth of highly anisotropic nanostructures during the thermal treatment. The high atomic mobilities attained at high temperatures favor nucleation and self assembly of the oxide material in the form of nanowires that grow along the less strained directions in order to minimize the elastic strain energy. The nanowires composition checked with TEM and EELS showed that they were also void of manganese. This result is as expected, since the growth principle is the same one used to grow nanodots.

Nanoimprint lithography

Our next objective was the preparation of nanostructured surfaces by direct nanoimprinting of PVOH based functional oxide precursor films. The majority of research in the field comes from work done on Polymethyl methacrylate (PMMA) or Polystyrene [10,11]. The advantage of imprinting directly into the PVOH based oxide precursors is the reduction in the number of steps required to obtain nanocrystal arrays, however this requires of a fine tuning of the imprint condition.

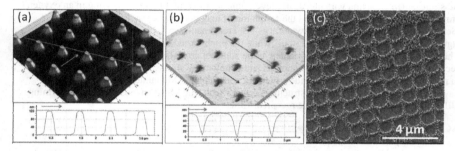

Figure 3. (a) AFM image of one of the typical molds. (b) Nanoimprinted precursor film on Si, dwell temperature 160 °C, applied pressure 40 MPa, dwell time 5 min, demolding temperature 80 °C, from nitrate based LSMO precursor with 2 wt% PVOH, (c) Final nanostructured surface after 900 °C annealing temperature for 240 minutes

The more specific procedure related to nanostructuring a LSMO precursor with a male mold by NIL is as follows [12-14]. First a LSMO film is spin coated on a single crystal substrate. The film is then heated above its glass transition temperature while in contact with the silicon mold. The mold is then pressed into the film and left there for a time scale of minutes. The mold is disengaged from the film once the temperature of the film is then lowered well below the glass transition temperature leaving a series of nanopores behind. The nanopores however do not reach to the substrate surface, thus a reactive ion etching step must be performed. This removes the excess material leaving smaller features than previous. The sample is finally annealed further reducing the size of the perforated film but causing the film to crystallize

Preliminary results of our imprinting approach are seen in figure 3. On figure 3(a) one can see one of our silicon molds. The pillars were spaced 1 µm, 125 nm high, 500 nm in width at the base and tapered very slightly towards their tips where the width was reduced to 250 nm. On figure 3(b) we observe the nice transfer of the mold motives into the precursor film after optimizing the imprint conditions. On figure 3(c) one can see the nanoholes array defined on top of a silicon substrate by NIL after RIE etching of the residual layer and a high temperature treatment. The nanoholes are defined in a polycrystalline LSMO matrix, since the substrate used was Si. The substrate floor is also seen there, thus signifying that the etching process was able to eliminate the residual layer.

TEM and EELS performed in NIL prepared samples showed that manganese was present in the correct ratio as per $La_{0.7}Sr_{0.3}MnO_3$. The result was different from what was observed in the nanowires and nanodots made from e-beam lithography. This suggests that the electron beam is somehow the cause of the manganese loss and future experiments using this procedure with oxides will need to take this into account.

CONCLUSIONS

Through the use of electron beam lithography, arrays of oxide nanostructures were generated on insulating single crystal STO, LAO and YSZ substrates. Specifically, nanodots and nanowires were written by spin coating a PVOH containing metal oxide precursor film on said substrates, then exposing with electrons, developing in water, and finally annealing in an oxygen atmosphere. The mechanism behind the formation of nanodots was cross-linking the polymer locally with focused electrons at low dosages. To create nanowires, the dosage was increased significantly with a lower limit of 150 pC while the pitch varied between 1000 nm and 250 nm. The nanowires were seen to grow oriented at the adjacent cross-linked precursor zone from secondary electrons. The resulting nanodots and nanowires were inspected with TEM and EELS to reveal single crystal epitaxial oxide structures with depleted Mn content. Additionally, we have shown the possibility to nanostructure surfaces with true LSMO nanostructures performing NIL in the same PVOH based oxide precursor films.

ACKNOWLEDGMENTS

We acknowledge the financial support from MICINN (MAT2008-01022, Consolider NANO-SELECT and GICSERV program 'Access to ICTS integrated nano- and microelectronics cleanroom') and Generalitat de Catalunya (Catalan Pla de Recerca 2009 SGR 770 and XaRMAE). J. Malowney acknowledges financial support from the AGAUR, Generalitat de Catalunya, grant resolution IUE/2681.

REFERENCES

1. S. Yamanaka, T. Kanki, T. Kawai, H. Tanaka, *Nano Lett.* 11, 343-347 (2011).
2. K.-Ch. Hyun *et al., Journal of Physical Chemistry C* 113, 7085-7090 (2009).
3. A. Carretero-Genevrier *et al., Advanced Functional Materials* 20, 892-897 (2010).
4. J. Zabaleta *et al., Journal of Applied Physics* 111, 024307 (2012).
5. A. Carretero-Genevrier *et al., Chem. Comm.* 48, 6223-6225 (2012).
6. J.M.D. Coey, M. Viret, S. von Molnar, *Adv. Phys.* 48, 167-293 (1999).
7. A.M. Haghiri-Gosnet, J.P. Renard, *J. Phys. D: Appl. Phys.* 36, R127-R150 (2003).
8. C.M. Chuang *et al., Nanotechnology* 17, 4399-4404 (2006).
9. M.C. Wu *et al., Journal of Materials Chemistry* 18, 780-785 (2008).
10. J. Tao *et al., Microelectronic Engineering* 78–79, 665-669 (2005).
11. C. Martin, L. Ressier, and J.P. Peyrade, *Physica E: Low-dimensional Systems and Nanostructures* 17, 523-525 (2003).
12. S.Y. Chou and P.R. Krauss, *Microelectronic Engineering* 35, 237-240 (1997).
13. L. Guo, P.R. Krauss, and S.Y. Chou, *Applied Physics Letters* 71, 1881-1883 (1997).
14. A. Lebib *et al., Microelectronic Engineering* 46, 319-322 (1999).

Mater. Res. Soc. Symp. Proc. Vol. 1547 © 2013 Materials Research Society
DOI: 10.1557/opl.2013.506

Melting Gel Films for Low Temperature Seals

Mihaela Jitianu[1], Andrei Jitianu[2*], Michael Stamper[1], Doreen Aboagye[2], Lisa C. Klein[3]

[1]Department of Chemistry, William Paterson University, 300 Pompton Road, Wayne, New Jersey 07470
[2]Department of Chemistry, Lehman College, CUNY, Davis Hall, 250 Bedford Park Boulevard West, Bronx, New York 10468
[3]Department of Materials Science and Engineering, Rutgers University, 607 Taylor Road, Piscataway, New Jersey 08854

ABSTRACT

Melting gels are silica-based hybrid gels with the curious behavior that they are rigid at room temperature, but soften around 110°C. A typical melting gel is prepared by mixing methyltriethoxysilane (MTES) and dimethyldiethoxysilane (DMDES). MTES has one methyl group substituted for an ethoxy, and DMDES has two substitutions. The methyl groups do not hydrolyze, which limits the network-forming capability of the precursors. To gain insight into the molecular structure of the melting gels, differential scanning calorimetry and oscillatory rheometry studies were performed on melting gels before consolidation. According to oscillatory rheometry, at room temperature, the gels behave as viscous fluids, with a viscous modulus, $G''(t,\omega_0)$ that is larger than the elastic modulus, $G'(t,\omega_0)$. As the temperature is decreased, gels continue to behave as viscous fluids, with both moduli increasing with decreasing temperature. At some point, the moduli cross over, and this temperature is recorded as the glass transition temperature T_g. The T_g values obtained from both methods are in excellent agreement. The T_g decreases from -0.3°C to -56°C with an increase in the amount of di-substituted siloxane (DMDES) from 30 to 50 mole %. A decrease of the T_g follows an increase of the number of hydrolytically stable groups, meaning a decrease in the number of oxygen bridges between siloxane chains.

INTRODUCTION

Hybrid materials are studied primarily for their ability to combine the properties of a high temperature inorganic material with low temperature processing. Silica-containing hybrid organic-inorganic gels can be used, for example, for films and coatings, due to the fact that the inorganic silica backbone provides hardness and stability, while the organic components provide flexibility, low temperature of processing and hydrophobicity [1,2]. The phenomenon of softening in hybrid silica-based gels first was reported by Matsuda et al. [3], who observed this behavior in the poly(benzylsilsesquioxane) system. He named these gels "melting gels". These hybrid gels differ from classical hybrids, or "ormosils", in that hybrid melting gels are solid at room temperature, become fluid at a modest temperature T_1 ($\sim$ 110°C), return to the rigid condition at room temperature, and can be cycled through softening and becoming rigid many times. However, after consolidation at a temperature T_2 ($T_2 > T_1$), the gels no longer soften. The consolidation temperature T_2 corresponds to cross-linking of the silica chains [4, 5]. Our previous studies investigated the preparation and properties of melting gels in methyl [5] and phenyl substituted siloxane gels [6].

Kakiuchida et al. [7, 8] investigated the viscoelastic properties in phenylpolysiloxane melting gel system. The flow of the phenylpolysiloxane molecules was related to the molecular volume and the intramolecular structure by using a free-volume model. They showed that the viscoelastic behavior is related to the number of bridging oxygens between silicon atoms and the distribution of the molecular weight [8]. Moreover, it was shown that an increase in the number of phenyl groups decreased the number of bridging oxygens, leaving only weak interactions between three-dimensional siloxane networks [7]. It also was shown that the organic groups influence directly the elastic modulus $G'(t, \omega_0)$, which can be controlled through the type and the number of organic groups bonded to the siloxane network [8]. The elastic modulus also shows a dependence on temperature. Masai et al. [9] observed that the temperature dependence of the elastic modulus did not change with heat treatment. This suggests that the phenyl modified system has a single relaxation process, which can be correlated with the structure of the gels. Finally, it was found that the type and the number of organic groups influence the glass transition temperature (T_g) [2,10]. An increase in the number of large pentyl or octyl groups, when condensation to form bridging oxygens is prevented, leads to lower T_g values [2]. In this study the thermal and rheological properties of the melting gels were investigated before their consolidation, by means of the determination of the glass transition temperature. The correlation of these data obtained through the two different methods is the key of this research.

EXPERIMENTAL

Gel preparation

The preparation of the melting gels was reported before [2, 3]. The melting gels were obtained by using a mono-substituted and a di-substituted alkoxysilane. The mono-substituted component was methyltriethoxysilane (MTES) (Sigma-Aldrich) while the di-substituted component was dimethyldiethoxysilane (DMDES) (Fluka Chemicals). The substituted alkoxides were used without further purification. Hydrochloric acid (Fisher Scientific) and ammonia (Sigma-Aldrich) were used as catalysts. Anhydrous ethanol (Sigma-Aldrich) was employed as solvent. Five gels were prepared with the indicated mol % of mono-substituted and di-substituted alkoxides listed in Table 1.

Table 1. Chemical compositions of the melting gels prepared in the MTES-DMDES system and their temperatures of consolidation along with the glass transition temperatures

Sample	MTES (mol %)	DMDES (mol %)	Consolidation Temperature (°C)	Glass Transition Temperature T_g (DSC) (°C)	Glass Transition Temperature T_g (Rheometry) (°C)	Activation energy (for η) E_a (kJ/mol)
1	75	25	135	-0.3	-0.15	212.6
2	70	30	140	-6.4	-6.7	205.2
3	65	35	150	-18.8	-20.6	194.4
4	60	40	155	-37.7	-27.2	183.0
5	50	50	160	-56.7	-50.4	133.8

The synthesis was performed in three steps. First the water was mixed with hydrochloric acid and half of the ethanol. The MTES was mixed with the other half of the ethanol separately. Then, the ethanol mixed with MTES was added dropwise to the water solution under continuous stirring. The container was covered tightly, and the mixture was stirred at room temperature for 3 hours. The molar ratios of MTES:EtOH:H_2O:HCl were 1:4:3:0.01.

In the second step, the di-substituted alkoxide DMDES was diluted with ethanol in a molar ratio of DMDES:EtOH = 1:4. The DMDES-EtOH mixture was added dropwise to the mixture from the first step. This solution was kept in a closed container under stirring for two more hours at room temperature.

In the third step, ammonia was added to the reaction mixture and the solution was stirred for another hour in a closed container. Then the clear solution was stirred for 48 hours at room temperature in an open container until gelation occurred. The gels were heat treated at 70°C overnight in order to remove excess ethanol. During this process, a white powder of ammonium chloride formed on the gels. The ammonium chloride was not soluble in the gel. To remove the ammonium chloride, 10 ml of acetone (Sigma-Aldrich) was added to the samples. The ammonium chloride was removed by vacuum filtration. The clear solution was then concentrated by stirring in order to remove the excess acetone followed by a heat treatment at 70°C for 24 hours to remove the solvents. A final heat treatment at 110°C was performed for removal of un-reacted water.

It was observed that after these heat treatments, the gels were solid at room temperature. However, when heated to ~110°C, the gels softened and became fluid. The consolidation temperature was established empirically by going through heating and cooling cycles until finding the temperature after which the gels could not be softened. Once the gel had been heated to the consolidation temperature, the softening behavior was no longer reversible. The consolidation temperatures are listed in Table 1. In this study the thermal and rheological behavior of the melting gels was investigated before their consolidation.

Materials characterization

The thermal behavior of the hybrid gels before consolidation was studied using differential scanning calorimetry (DSC TA-Q-2000). The DSC analyses were recorded using a 5°C/min heating rate between -75 and 400°C under nitrogen flow. The viscosity at low temperatures and the rheological properties were investigated using oscillatory rheometry (TA AR-G2, using a 3°C/min heating rate in air) on melting gels before their consolidation. The temperature sweep oscillatory measurements were performed from room temperature to -100°C, with a temperature ramp of -3°C/min at a constant stress of 100Pa. The oscillatory frequency was chosen according to the linear viscoelastic domain for each gel. This was determined by performing frequency sweep tests prior to the temperature sweep tests. For samples 75%MTES - 25%DMDES and 70%MTES - 30%DMDES, frequency ω_0 has been set at 6.283 rad/s, while for the remaining three samples, ω_0 has been set at 0.6283 rad/s. When the glass transition temperature was reached for each gel, the temperature sweep test was stopped.

The complex viscosity (η) was estimated from the values of viscous modulus, G" and the angular frequency, ω_0 ($\eta = G"/\omega_0$). The temperature dependence for the viscosity was evaluated assuming that viscosity is an activated process, according to an Arrhenius model ($\ln\eta = A + E_a/RT$, where A is a preexponential factor, E_a is the activation energy to reach the vitreous state, and R is the ideal gas constant, 8.3145 J/mol K).

83

RESULTS AND DISCUSSION

MTES is a mono-substituted alkoxide, while DMDES is a di-substituted akoxide. The MTES has three ethoxy functional groups which can be hydrolyzed and which can be further subject to polycondensation, while DMDES has only two ethoxy functional groups which can be further subject to polycondensation. The methyl groups (CH_3) are blocking one functionality for MTES and two for DMDES. This has a consequence in the cross linking processes. For example, the use of DMDES alone leads to poly-dimethylsiloxane (PDMS) which contains only linear polymeric chains. Hence PDMS is a liquid at room temperature. The melting gels studied here were obtained through co-condensation of MTES and DMDES. When a mono-substituted alkoxide is mixed with a di-substituted alkoxide, the di-substituted alkoxide act as a bridge between the molecular species formed when the mono-substituted alkoxide hydrolyzes and undergoes condensation polymerization. An increase in the ratio of DMDES to MTES decreases the tri-dimensional cross-linking, decreasing the oxygen bridges between the siloxane chains. This structural modification of the melting gels results in changes of the rheological properties and also of the glass transition temperature.

In this study, the glass transition temperature (T_g) values were evaluated using both Differential Scanning Calorimetry (DSC) and oscillatory rheometry measurements. The DSC curves and the temperature sweep plots of samples 75%MTES – 25%DMDES and 60%MTES – 40%DMDES are presented in Figures 1 and 2. In the DSC scans, the inflection points are identified with the glass transition behavior, and these values, along with the values from the oscillatory rheometry measurements are listed in Table 1. The T_g values show a linear variation and a continuous decrease with an increase in the concentration of DMDES.

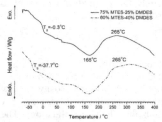

Figure 1. The differential scanning calorimetry (DSC) curves for melting gels which contain 75%MTES - 25% DMDES and 60%MTES - 40% DMDES

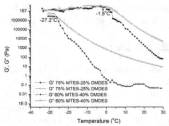

Figure 2. The temperature sweep plots for melting gels which contain 75%MTES - 25% DMDES and 60%MTES - 40% DMDES

Oscillatory rheometry tests allow direct determination of the glass transition temperature, as shown in Figure 2. The elastic modulus $G'(t,\omega_0)$ and viscous modulus $G''(t,\omega_0)$ were measured in small amplitude oscillatory shear at constant angular frequency ω_0 as a function of decreasing temperature – temperature sweep tests. The crossover point at low temperature identified on Figure 2 has been assigned to the glass transition temperature T_g. At room temperature, the gels behaved as viscous fluids, with viscous modulus, G" being larger than the elastic modulus, G'. As the temperature decreased, gels continued to behave as viscous fluids, with both moduli increasing with decreasing temperature, until they cross each other.

There is little difference between the T_g values obtained by both methods, although for samples 60%MTES – 40%DMDES and 50%MTES – 50%DMDES the difference was slightly larger. These samples contain the higher concentrations of DMDES, and these tend to have lower viscosity values. The free volume theory [11] governs the glass transition phenomenon. The T_g is associated with the translational mobility of the molecules, which effects the potential energy by changing the interatomic and intermolecular spacing between neighboring molecules, leading to changes in G' and G" values by a factor of one thousand over a small temperature range. Thus, a higher concentration of DMDES has an influence on the configuration of the chains and therefore on the intramolecular potential energies. In addition, the increased number of methyl groups on the polymeric chains can increase the free volume, which leads to slightly lower glass transition temperatures.

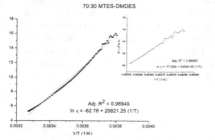

Figure 3. The lnη as a function of reciprocal temperature, along with data fit accounting for an Arrhenius model temperature dependence for sample 70%MTES - 30% DMDES (Insert – restricted temperature fit)

Figure 4. The variation of E_a (filled circle) and T_g (open circle) obtained from oscillatory measurements as a function of temperature

The variations of the lnη versus 1/T were recorded for all samples. Figure 3 displays this variation for sample 70%MTES - 30% DMDES. The data have been fit according to an Arrhenius model for temperature dependence, the variation of lnη versus reciprocal of temperature (1/T). While it is possible to extract an activation energy from the plot, it is clear that the data do not fit an Arrhenius model over the entire viscosity range, especially for the gels with lower viscosity. Even in systems where it is expected that the viscosity shows Arrhenius behavior [12], it is difficult to fit to a single activation energy over a wide range of temperatures. An alternative approach is to fit the data were over a restricted temperature range, more towards lower temperatures, approaching T_g. A fit closer to an Arrhenius model was found over this restricted temperature range.

The calculated values for the activation energy (E_a) are given in Table 1. The values for E_a appear to follow the same trend as the glass transition temperature. The T_g decreases steadily with an increase of DMDES in the gels, as displayed in Figure 4.

The glass transition is generally thought of as the interval of transition between liquid states and glassy. Thus, T_g is an indication of the degree of cross-linking in the silica network. We can conclude that by increasing the concentration of the DMDES, the concentration of the methyl groups is increased, which leads to a decrease in the number of oxygen bridges between silicon atoms. What is interesting about the melting gels is that the T_g values are below room temperature going down to as low as -50°C. Low T_g polymers are used in imprint lithography, where the polymer is molded above T_g and demolded below T_g in order to transfer patterns with high resolution [13]. Judging from the behavior found in the MTES-DMDES melting gels, the low T_g can be used in the same way to extend imprint lithography to hybrid materials.

CONCLUSIONS

The preparation of five compositions of melting gels using a mono-substituted alkoxide (MTES) along with a di-substituted alkoxide (DMDES) was investigated. The thermal and rheological properties of the melting gels before their consolidation into hybrid glasses were explored. The glass transition temperatures (T_g) were measured using Differential Scanning Calorimetry (DSC) and oscillatory rheometry. It was observed that T_g decreases from -0.3°C to -56.7°C from DSC analysis, while in the temperature sweep experiments the T_g decreases from -0.15°C to -50.4°C. Thus, for most samples, the values for T_g found by both techniques are very close. The T_g decreased when the concentration of DMDES increased, which is due to the decreasing of the number of the bridging oxygens between the silicon atoms. The number of bridging oxygens is limited by an increase in the number of the methyl groups that cannot hydrolyze. The activation energies (E_a) were determined from the variation of lnη as a function of reciprocal temperature. The E_a decreased when the concentration of DMDES increased, in parallel with the decrease in T_g with the increase in the number of methyl groups.

REFERENCE

[1]   A. Kamer, K. Larson-Smith, L.S.C. Pingree, R.H. Dauskardt, Thin Solid Films, **519**, 1907-1913, (2011).
[2]   J. Macan, T. Tadanaga, M. Tatsumisago, J. Sol-Gel Sci. Technol, **53**, 31-37, (2010).
[3]   A. Matsuda, T. Sasaki, K. Hasegawa, M. Tatsumisago and T. Minami, J. Am. Ceram. Soc. **84**, 775-780, (2001).
[4]   A. Jitianu, J. Doyle, G. Amatucci, L. C. Klein, "Methyl-modified melting gels for hermetic barrier coatings", Proceedings MS&T 2008 *Enabling Surface Coating Systems: Multifunctional Coatings* (CDROM), Pittsburgh, PA, 2008, pp. 2171-2182.
[5]   A. Jitianu, J.P. Doyle, G. Ammatuci, L. C. Klein, , J. Sol-Gel Sci. Technol, **53**, 272-279, (2010).
[6]   A. Jitianu, G. Amatucci, L. C. Klein, J. Am. Ceram. Soc. **92**, 36-40, (2009).
[7]   H. Kakiuchida, M. Takahashi, Y. Tokuda, H. Masai, M. Kuniyoshi, T. Yoko, J. Phys. Chem. B, **110**, 7321-7327, (2006).
[8]   H. Kakiuchida, M. Takahashi, Y. Tokuda, H. Masai, M. Kuniyoshi, T. Yoko, J. Phys. Chem. B **111**, 982-988, (2007).
[9]   H. Masai, Y. Tokuda and T. Yoko, J. Mat. Res., **20**, 1234-1241, (2005).
[10]   A. Jitianu, K. Lammers, G.A. Arbuckle-Kiel, L.C. Klein J. Therm. Anal. Cal., **107**, 1039-1045, (2012).
[11]   C. L. Rohn, Analytical Polymer Rheology, Hanser Publishers, Vienna, New York, 1995, pages 171-174.
[12]   D. Giordano, M. Potuzak, C. Romano, D.B. Dingwell, M. Nowak, Chemical Geology **256**, 203-215, (2008).
[13]   Y. Hirai, K. Kanakugi, T. Yamaguchi, K. Yao, S. Kitagawa, Y. Tanaka, Microelectronic Eng. **67-68**, 237-244, (2003).

Mater. Res. Soc. Symp. Proc. Vol. 1547 © 2013 Materials Research Society
DOI: 10.1557/opl.2013.578

Low Temperature Syntheses of Transition Metal Bronzes with an Open Structure for High Rate Energy Storage

X. Pétrissans[1], V. Augustyn[2], D. Giaume[1], P. Barboux[1], B. Dunn[2]

[1] Laboratoire de Chimie de la Matière Condensée, Chimie-ParisTech, 11 rue Pierre et Marie Curie, 75005 Paris, France

[2] Department of Materials Science and Engineering, University of California Los Angeles, Los Angeles, CA 90095, USA

ABSTRACT

Development of devices storing and delivering high-energy power such as supercapacitors is necessary to assist intermittent sources of energy. Most of the commercial systems are carbon-based, but due to their high surface charge, oxides offer a valuable alternative for high-rate energy storage. Among them, layered transition metal oxides with mixed valence properties present both good electronic and ionic conductivities suitable for application to electrochemical applications intermediate between capacitors and batteries. This work focuses on lamellar oxide bronzes based on cobalt M_xCoO_2 and vanadium $M_xV_2O_5$ (M = H, Li, Na or K). A low temperature synthesis leads to high specific area particles (above 100 m^2/g). Hydrated and anhydrous Na_xCoO_2 are promising cathode materials for aqueous supercapacitors, with a high capacity of more than 100 mAh/g obtained under 20 mV/s for the hydrated Na_xCoO_2. The $M_xV_2O_5$ bronzes appear to be good candidates for organic supercapacitors, especially the $Li_xV_2O_5$ bronze, which shows a high stable capacity above 100 mAh/g (at 20 mV/s ie a charging time of 125 s).

INTRODUCTION

Supercapacitors are electrochemical devices that can deliver higher electrical power than conventional batteries. Their properties are based on the accumulation of charges at the interface between a solid and the electrolyte [1]. There is no chemical reaction or solid-state diffusion limitations resulting in a low internal resistance and high current densities. However, the amount of surface charges at the solid-electrolyte interface is limited and therefore limits the energy density (5-10 Wh/kg in commercial carbon-based systems). Due to their high surface charge, oxides offer a valuable alternative to carbon-based electrochemical capacitors for high-rate energy storage [2]. In addition to conventional charge accumulation at the double layer interface, transition metal oxides with mixed valence properties allow redox reactions in their outer layers that largely increase the capacitance value (known as pseudo-capacitance [3]). The best example is that obtained for the expensive ruthenium oxide [4]. In addition, there is a continuous transition form capacitive to faradaic processes in the bulk of the materials with high specific area such as thin films or nanoparticles. This effect can be further increased in the case of open-structure materials which present a good ionic conductivity and fast ionic intercalation and exchange properties. This work will particularly focus on cobalt oxides and vanadium bronzes presenting a lamellar-structure. Indeed, such structures can achieve good ionic mobility with good cycling reversibility [5, 6].

In this paper, we particularly focus on layered cobalt oxides especially on the $Na_{0.6}CoO_2$ phase [7, 8]. The good electronic conductivity of this material associated to its fast ion exchange properties makes it a good candidate for high rate energy storage [9].

The vanadium bronzes $M_xV_2O_5$ (M=Li, Na, K) are also known to be better electronic conductors than most oxides and their 2D lamellar structure allows a good ionic mobility and an easy insertion of alkaline ions [10].

EXPERIMENTS

1) Cobalt bronzes synthesis

Nanoparticles of cobalt oxides have been synthesized by direct precipitation following a method described elsewhere [11]. Typically, in a three neck round bottom flask, 22 g of NaOH are dissolved in 50 mL of deionized water under oxygen flow. Then 0.7 mL of Br_2 are slowly added leading to a yellow-orange solution. Immediately afterwards a 1 mol/L $Co(NO_3)_2$ is added dropwise. The solution turns black and the precipitate is then recovered through vacuum filtration. The obtained black paste is dried at 120°C during 24 h. The minimum amount of deionized water is then used to remove the remaining NaOH. After another drying step at 120°C for 24 h and grinding, a fine $Na_{0.6}CoO_2$ black powder is obtained.

2) Cobalt bronzes electrochemical measurements

Following the previously published method by Wang et al. [2], the electrodes were made by depositing the $Na_{0.6}CoO_2$ powder onto a glassy carbon substrate. The main interest of this experiment is to get rid of the effects of carbon additives and/or polymeric binder and to avoid polarization limitations within the electrode. First, a 1 cm x 2 cm glassy carbon piece is plasma-treated during 15 min in oxygen to allow the surface to become hydrophilic. Then approximately 10 µL of a colloidal dispersion of the powder in a 10 mol/L NaOH solution (2 g/L) is drop-casted onto the treated surface. To allow good contact between the particles and the substrate, the electrodes were further heated up during 40 min at 220°C for the $Na_{0.6}CoO_2.yH_2O$, and at 300°C for the $Na_{0.6}CoO_2$.

A standard three electrode setup has been used in a 10 M NaOH solution as electrolyte. An Ag/AgCl reference electrode and a carbon sheet counter electrode were used. Moreover, during the experiment, Teflon tape covers the other side of the glassy carbon substrate to avoid any leakage current.

3) Vanadium bronzes synthesis

Nanoribbons of the red brownish $V_2O_5.xH_2O$ gel are obtained thanks to a sol-gel synthesis starting from a sodium metavanadate Na_3VO_4 solution [12, 13]. Sodium is exchanged for protons through a strong acidic ion exchange resin Dowex 50WX2 and the resulting reaction may be summarized as the condensation of polyvanadic acids to an hydrated form of vanadium oxide with a composition of $V_2O_5.1.6\ H_2O$ ($P_{H2O} = 10$ mmHg, 20°C) in equilibrium with the relative pressure of water [13].

$$Na_3VO_4 \xrightarrow{H^+} H_3VO_4 \longrightarrow H_6V_{10}O_{28} \longrightarrow H_{2x}V_2O_{5+x} \approx V_2O_5.1.6\ H_2O$$

Then intercalation of alkaline ions M (M = Li^+, Na^+, K^+) is obtained either in solutions of alkali iodides (with redox insertion reaction and formation of a mixed valent V^V-V^{IV} oxide) following reactions (1), or of alkali sulfates (limited to simple ionic exchange) following reaction (2).

Alkaline ion bronzes obtained by insertion and oxidation of iodide

Reduction step: $y\ MI + H_{2x}V_2O_{5+x} \rightarrow H_{2x}M_yV_2O_{5+x} + y/2\ I_2$

Ion exchange step: $2x\ MI + H_{2x}M_yV_2O_{5+x} \rightarrow M_{y+2x}V_2O_{5+x} + 2x\ HI$ (green)   (1)

Ion exchange (obtained with sulfate as the counterions)

$x\ M_2SO_4 + H_{2x}V_2O_{5+x} \rightarrow M_{2x}V_2O_{5+x} + x\ H_2SO_4$ (red)   (2)

To achieve these complete intercalations, the V_2O_5 gel is directly soaked into the previously mentioned solutions (0.1 mol/L) during one week. Then the precipitates are washed with deionized water to remove the remaining iodide or sulfate and further dried at 120°C for 24 hours.

4) Vanadium bronzes electrochemical measurements

A slurry is first made of $M_xV_2O_5$ mixed with 15 wt% of acetylene black and 5 wt% of PVdF previously dissolved in N-methyl-pyrolidone. To this slurry, 20 vol% of dibutylphtalate (DBP) added to obtain a paste, which is then cast thanks to a Doctor Blade® set to a thickness of 50 μm onto a cleaned aluminum foil (200 μm thick). The electrode is dried then at 120°C for 4 h and further pressed at 740 MPa at 120°C. Finally, DBP is removed in diethylether to form porosity in the electrode.

Electrochemical experiments were carried out in a two electrode Swagelok® cell using a lithium foil (or sodium foil) as the counter and reference electrode, in a 1 mol/L $LiClO_4$ (or 1 mol/L $NaClO_4$) propylene carbonate electrolyte: the pristine vanadium oxide, $Li_xV_2O_5$ and $K_xV_2O_5$ used the lithium based cell while $Na_xV_2O_5$ used the sodium based cell.

RESULTS AND DISCUSSION

1) Cobalt bronzes

Nanoparticles of $Na_{0.6}CoO_2$ present a high specific area (above 100 m²/g). As shown in Figure 1, particles exhibit a platelet shape. For the hydrated compound (a) the particles are 5 nm wide and 10 to 20 nm long packed in some amorphous layer, whereas the dried phase (b) presents well dispersed particles of 6 nm wide and 30 nm long. Moreover, the main difference between these two phases is the interlamellar space (d_{001}). For instance the hydrated one has an interlamellar space of 6.6 Å whereas the dried one exhibits a smaller interlamellar space of 5.5 Å, consistent with the previously published results on $Na_{0.6}CoO_2$ [14].

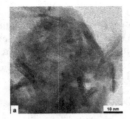 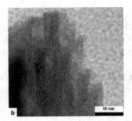

Figure 1: Transmission Electron Microscopy pictures of (a) $Na_{0.6}CoO_2.yH_2O$ and (b) anhydrous $Na_{0.6}CoO_2$.

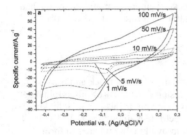 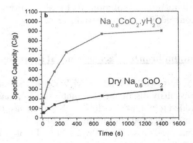

Figure 2: (a) Third voltammograms of $Na_{0.6}CoO_2.yH_2O$ in 10 mol/L NaOH at 1, 5, 10, 50, 100 mV/s, (b) Specific capacity of $Na_{0.6}CoO_2.yH_2O$ and dry $Na_{0.6}CoO_2$ as a function of the charging time.

As shown in Figure 2 (b), for $Na_xCoO_2.yH_2O$ deposited onto a glassy carbon substrate in a 10 mol/L NaOH electrolyte, at slow cycling rates (1 mV/s, charging time of 1400 s), high capacity of 900 C/g is measured. This high capacity can be attributed to the reversible faradaic insertion of 0.5 sodium/Co in addition to the double layer capacitance. But the dry phase exhibits much lower capacity probably because ionic diffusion is limited.

Furthermore, the anodic peak intensities observed on the voltammograms of Figure 2(a) characterize the materials kinetics. If the potential variation is modified, the peak intensity is also varying, following the equation proposed by Lindström et al. [15]:

$$I_{peak} = av^b$$

where the b factor has a value comprised between 0.5 and 1. If the limitation kinetics is the mass transport, then b=0.5. When the charge transfer is the limitation step, b=1. To summarize, b=0.5 means that the material has a faradaic behavior, whereas b =1 is characteristic of a capacitive behavior.

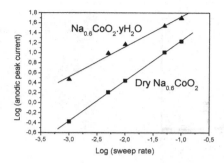

Figure 3: Logarithm of the anodic peak current of both the hydrated and the dry $Na_{0.6}CoO_2$ as a function of the logarithm of the sweep rate.

As shown in Figure 3, the slope of the hydrated phase is lower than the slope of the dry compound. The b factors are shown in the Table I.

Table I: Summary for the hydrated and anhydrous cobalt bronzes showing the b factor, the interlamellar space and the specific capacity/capacitance at 20 mV/s.

Compound	b	d_{001} (Å)	Specific Capacity at 20 mV/s in C/g (mAh/g)
$Na_{0.6}CoO_2.yH_2O$	0.61	6.6	360 (100)
$Na_{0.6}CoO_2$	0.83	5.5	90 (25)

The hydrated phase of the cobalt bronzes exhibits a lower b factor than the dry one, meaning that it has a more faradaic behavior. This can be explained by the larger interlamellar space which allows faster diffusion of ions. As a result, specific capacities measured at 20 mV/s are quite different: the specific capacity of the hydrated bronze is four times higher than for the dry one at a quite high sweep rate.

2) Vanadium bronzes

Nanoribbons of hydrous vanadium oxide ($H_{2x}V_2O_{5+x}$ x~0.5) are obtained. As shown in the Figure 4 (b) ribbons are 10 to 20 nm wide and 1 µm long.
The diffraction diagram has been made on a previously heated (120°C) film of the hydrous vanadium oxide. The diagram (Figure 4 (a)) shows that V_2O_5 is a lamellar compound. To investigate if insertion and/or reduction occurred, XRD experiments have been carried out: using the (003) reflection, the interlamellar space can be calculated using Bragg's law. The results of this study are summarized in Figure 5

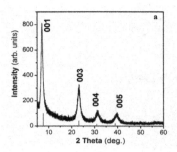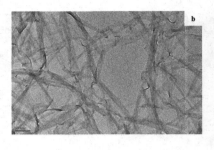

Figure 4: (a) X-ray diffraction diagram of a dry film of V_2O_5, (b) Transmission Electron Microscopy picture of the V_2O_5 gel.

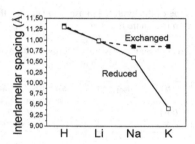

Figure 5: Interlamellar spacing of both the ion exchanged (■) and the reduced (bronze) ( ) vanadium oxides as function of the ion (Li^+, Na^+, K^+).

As shown in Figure 5 regarding the vanadium oxides, when the intercalation is performed by ion exchange, we hardly observe an effect on the interlamellar spacing. On the contrary, for bronzes obtained by reduction, we can clearly see a decrease in the interlamellar spacing with the size of the inserted cation. The reason is supposed to be the interlamellar water amount present in the case of small polarizing ions, which decreases when the ion is getting larger because its hydration energy is lower. This is especially true for potassium cations. In the bronzes, the charge of the vanadium oxide layer becomes negative with the reduction of V^V into V^{IV}. Hence the attraction between the negative oxide layer and the inserted cation leads to a smaller interlamellar spacing especially when it overcomes the hydration energy (case of potassium).

When cycled in $LiClO_4$ versus Li, all exchanged materials (Li, Na, K) give an electrochemical behavior similar to the pristine hydrated V_2O_5 material with a capacity of 90 mAh/g at 1 mV/s (Figure 6). This is not surprising because they probably all exchanged for lithium in contact with the electrolyte. However in the case of bronzes (reduced samples), larger differences are observed as shown in Figure 6.

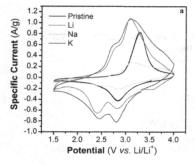 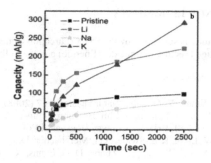

Figure 6: (a) Third voltammograms of the pristine V_2O_5 and the vanadium bronzes ($Li_xV_2O_5$, $Na_xV_2O_5$, $K_xV_2O_5$) at 1 mV/s, (b) Measured capacity of the pristine V_2O_5 and the vanadium bronzes ($Li_xV_2O_5$, $Na_xV_2O_5$, $K_xV_2O_5$) as a function of the charging time.

As shown in the Figure 6(a) presenting the voltammograms of the pristine and bronzes, the shape of the bronzes voltammograms is clearly different than the pristine vanadium oxide one. Even though $K_xV_2O_5$ seems to offer the best capacity at a slow sweep rate (Figure 6(b)), this capacity is not stable. Indeed such high capacity is due to an ionic exchange between Li^+ (present in the electrolyte) and K^+ present in the structure. Thus, the cycling reversibility of the $K_xV_2O_5$ (not shown here) is rather poor. But the $Li_xV_2O_5$ bronze shows a high stable capacity above 200 mAh/g (charging time 2500 s). And even at a high sweep rate (20 mV/s, charging time 125 s), the capacity remains above 100 mAh/g proving a fast ionic conductivity as well as a good electronic conductivity. Moreover $Na_xV_2O_5$ has a lower capacity whatever the sweep rate. This may be explained by an interlamellar spacing smaller than for $Li_xV_2O_5$ and perhaps a lower ionic diffusion of sodium in the V_2O_5 framework.

CONCLUSIONS

Low temperature synthesis of intercalation materials based on cobalt or vanadium ions have been successfully performed, leading to high specific area particles (above 100 m^2/g). Hydrated and dry Na_xCoO_2 revealed to be quite performant cathodic materials for aqueous systems, and high capacity of more than 360 C/g (100 mAh/g) under 20 mV/s has been reached for the hydrated Na_xCoO_2. $Li_xV_2O_5$ bronze shows the highest stable capacity of all vanadium bronzes, above 720 C/g (200 mAh/g) at 1 mV/s (charging time 2500 s). And even at a high sweep rate (20 mV/s, charging time 125 s), the capacity remains above 100 mAh/g proving both fast ionic and electronic conductivity. The reduction of the material prior to the electrochemical cycling seems to have a strong effect on the properties whereas simple exchange has no effect. In order to better understand the role of the conductivities on the electrochemical properties, further work has to be carried on.

ACKNOWLEDGMENTS

This study is supported by the DGA (Direction Générale de l'Armement), the RS2E (Réseau de Stockage Electrochimique de l'Energie) and the Fulbright Program.

REFERENCES

[1] P. Simon, Y. Gogotsi, Nature Materials 7 (2008) 845-854.

[2] J. Wang, J. Polleux, J. Lim, B. Dunn, J. Phys. Chem. C 111 (2007) 14925.

[3] W. G. Pell, Brian E. Conway, Journal of Power Sources 136 (2004) 334.

[4] J.W. Long, K.E. Swider, C.I. Merzbacher, D.R. Rolison, Langmuir 15 (1999) 780.

[5] Y. Wang, W. Yang, S. Zhang, D. G. Evans, X. Duan, J. Electrochem. Soc. 152 (2005) A2130.

[6] V. Gupta, S. Gupta, N. Miura, Journal of Power Sources 189 (2009) 1292.

[7] M. Pollet, M. Blangero, J. P. Doumerc, R. Decourt, D. Carlier, C. Denage and C. Delmas, Inorg. Chem. 48 (2009) 9671.

[8] K. Takada, H. Sakurai, E. Takayama-Muromachi, F. Izumi, R. A. Dilanian and T. Sasaki, Nature 422 (2003) 53.

[9] F. Tronel, L. Guerlou-Demourgues, M. Basterreix, C. Delmas, Journal of Power Sources 158 (2006) 722.

[10] I.D. Raistrick, R.A. Huggins, Solid State Ionics 9-10 (1983) 425-430.

[11] X. Pétrissans, A. Bétard, D. Giaume, P. Barboux, B. Dunn, L. Sicard, J.-Y. Piquemal, Electrochimica Acta 66 (2012) 306-312.

[12] N. Gharbi, C. Sanchez, J. Livage, J. Lemerle, L. Nejem, J. Lefebvre, Inorg. Chem. 21 (1982) 2158-2165.

[13] J. Livage, Coordination Chemistry Reviews, 178–180 (1998) 999–1018.

[14] M. Butel, L. Gautier, C. Delmas, Solid State Ionics 122 (1999) 271.

[15] H. Lindström, S. Södergren, A. Solbrand, H. Rensmo, J. Hjelm, A. Hagfeldt, S.-E. Lindquist, J. Phys. Chem. B, 1997, 101, 7717-7722.

Mater. Res. Soc. Symp. Proc. Vol. 1547 © 2013 Materials Research Society
DOI: 10.1557/opl.2013.782

Field Dependent Electrical Conduction in Metal-Insulator-Metal Devices using Alumina-Silicone Nanolaminate Dielectrics

Santosh K. Sahoo,[1,2] Rakhi P. Patel,[3] and Colin A. Wolden[3]

[1]National Renewable Energy Laboratory, 1617 Cole Blvd., Golden, Colorado 80401, USA

[2]New Jersey Institute of Technology, Newark, New Jersey 07102, USA

[3]Department of Chemical and Biological Engineering, Colorado School of Mines, Golden, Colorado 80401, USA

ABSTRACT

Hybrid alumina-silicone nanolaminate films were synthesized by plasma enhanced chemical vapor deposition (PECVD) process. PECVD allows digital control over nanolaminate construction, and may be performed at low temperature for compatibility with flexible substrates. These materials are being considered as dielectrics for application such as capacitors in thin film transistors and memory devices. In this work, we present the temperature dependent current versus voltage (I-V) measurements of the nanolaminate dielectrics in the range of 200- 310 K to better asses their potential in these applications. Various models are used to know the different conduction mechanisms contributing to the leakage current in these nanolaminate films. It is observed that space charge limited current (SCLC) mechanism is the dominant conduction process in the high field region whereas Ohmic conduction process is contributing to the leakage current in the low field region. The shallow electron trap level energy (E_t) of 0.16 eV is responsible for SCLC mechanism whereas for Ohmic conduction process the activation energy (E_a) for electrons is about 0.22 eV. An energy band diagram is given to explain the dominance of various conduction mechanisms in different field regions in these nanolaminate films.

INTRODUCTION

Hybrid nanolaminates comprised of polymer and inorganic layers are being pursued to serve a number of roles to enable flexible opto-electronics [1-3]. Thin film transistors (TFTs) are an important component in low cost, large area electronics such as flat panel displays and smart cards. The quality of the dielectric layer can significantly impact TFT performance, particularly when using organic semiconductors [4-6]. Ideally these materials provide a high specific capacitance to reduce turn-on voltage requirements (< 5 V), as well as low leakage to maintain a high on/off current ratios (>10^5). These demands have been met by inorganic nanolaminates, [7-9] but the use of these structures is limited in applications requiring flexibility. Multi-layer structures consisting of alternating inorganic and organic layers have been demonstrated to be effective dielectrics for devices fabricated on flexible substrates [1].

Various inorganic/organic hybrid dielectrics are explored to date and numerous material combinations have been explored [4-6,10-12]. Popular polymers include poly (4-vinylphenol) (PVP) and poly (methyl methacrylate) (PMMA). The inorganic material is typically a high dielectric constant (high-k) metal oxide (e. g. TiO_2, Ta_2O_5, Al_2O_3, HfO_2, YO_x). The different geometries studied so far include simple bilayers, polymer/oxide/polymer sandwich structures, and nanolaminates. These multilayered structures showed better leakage current and also improved dielectric properties in comparison to single layer thin films [13, 14].

The study of leakage mechanisms in nanolaminate dielectrics is very important in order to further improve the device performance. Different conduction mechanisms contributing to the leakage current in multilayered dielectric are studied such as Schottky mechanism, Poole-Frenkel, space charge limited current (SCLC), and ohmic conduction [15,16]. Detailed studies of the leakage mechanism has been reported for several inorganic-based nanolaminates [15,17]. However, to date there have been no reports that have examined the field dependent electrical conduction mechanism(s) in hybrid nanolaminates comprised of alternating polymer and oxide thin films. We recently introduced hybrid nanolaminates comprised of alumina and silicone [18]. Both films are deposited by plasma-enhanced chemical vapor deposition (PECVD) process in a single chamber. PECVD allows reductions in the thickness of the polymer layer relative to spin coating, and high specific capacitance values (>20 nF/cm^2) were achieved. The leakage current density at a field strength of 1 MV/cm was ~10^{-9} A/cm^2, while breakdown required applied electric fields in excess of 5 MV/cm [18]. In this work, we have studied the field dependent conduction mechanisms that contributes to the leakage current in these high performance alumina/silicone nanolaminates in different field regions. Temperature-dependent current versus voltage (I-V) measurements are taken in the temperature range of 200 to 310 K for these nanolaminates and single layer alumina films and employed to investigate the potential conduction mechanisms that are dominant in different field regions.

EXPERIMENTAL

Alumina/silicone nanolaminates and pure alumina single layer films were deposited on fluorine doped tin oxide (FTO, TEC-15) coated glass slides by plasma enhanced chemical vapor deposition (PECVD) process. The reactor used to deposit the alumina/silicone nanolaminates and alumina films in this work is a parallel plate, capacitively coupled PECVD system that has been described in detail previously [19]. The precursors for the alumina and silicone layers were trimethyl aluminum (TMA, Al(CH$_3$)$_3$) and hexamethyl disiloxane (HMDSO), respectively. The detail procedure for the deposition of individual alumina and silicone layer is given elsewhere [20]. For the nanolaminates described in this work the dyad thickness was fixed at 30 nm, and the alumina volume fraction was kept at 50 %. The nanolaminate was an 11-layer structure comprised of 5 dyads along with an extra Al$_2$O$_3$ layer [18]. Metal-insulator-metal (MIM) capacitors were fabricated by vacuum evaporation of aluminum contacts through shadow masks with a diameter of 840 μm on the nanolaminate films. The temperature dependent I-V measurements were carried out using an HP 4140B pA meter and cryostat.

RESULTS AND DISCUSSION

Figure 1(a) shows the leakage current density (*J*) versus electric field (*E*) for a nanolaminate having 50% Al$_2$O$_3$ composition for the temperature range of 200 – 310 K. It is observed that *J* increases with the increase of temperature. The strong temperature dependence of the leakage current density (Fig.1(a)) suggests that either Schottky-barrier mechanism or Poole-Frenkel conduction is contributing to the leakage current. It is well known that the Poole-Frenkel mechanism is due to the field enhanced thermal excitation of trapped electrons into the conduction band. Defects below the conduction band within the dielectric bandgap can participate in the conduction process through electron trapping and detrapping when both field and temperature are varied. The leakage current density due to Poole-Frenkel mechanism is described by [21]

$$J_{PF} = qN_c \mu E \exp\left(\frac{\beta_{PF} E^{1/2} - q\phi_t}{kT}\right) \tag{1}$$

where $\beta_{PF} = (q^3/\pi\varepsilon\varepsilon_0)^{1/2}$ (ε is the dynamic dielectric constant and ε_0 is the permittivity of the free space), N_c is the density of states in the conduction band, μ is the electron mobility, $q\phi_t$ is the trap level energy, E is the electric field, k is the Boltzmann constant, and T is the device temperature.

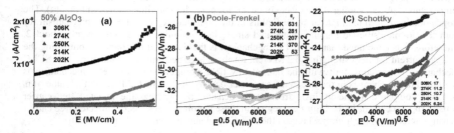

Figure 1. (a) J vs. E, (b) ln (J/E) vs. $E^{0.5}$, and (c) ln (J/T^2) vs. $E^{0.5}$ at different temperatures for alumina-silicone nanolaminate with 50% Al$_2$O$_3$.

The potential contribution of the Poole-Frenkel conduction mechanism is studied by plotting *ln (J/E)* versus $E^{0.5}$ as shown in Fig. 1(b). It is observed that a straight line fit is obtained in the high field region (> 0.3 MV/cm) for the temperatures range of 200 K to 310 K. The dynamic dielectric constant ε can be extracted from the slope of the linear fit in the high field region. The refractive index, n, of alumina-silicone nanolaminate film is about 1.6 [18] so that the expected value of $\varepsilon \sim n^2 = 2.56$ [20]. The estimated values of ε from the Poole-Frenkel plots come out to be very high in the range of 50 - 500. Therefore, the Poole-Frenkel mechanism is not contributing to the leakage current in the high field region for the nanolaminate samples.

Next, we explored the possibility that the Schottky mechanism may be contributing to the leakage current. The conduction current due to Schottky mechanism is described using the following relation [21]

$$J_{SC} = AT^2 \exp\left(\frac{\beta_{SC} E^{1/2} - q\Phi_b}{kT}\right) \tag{2}$$

where A is the effective Richardson constant, $q\Phi_b$ is the Schottky barrier height, and $\beta_{SC} = (q^3/4\pi\varepsilon\varepsilon_0)^{1/2}$, and the other terms are as described in Eq. 1. To evaluate the contribution of the Schottky mechanism to electrical leakage the I-V data is plotted as *ln (J/T^2)* versus $E^{0.5}$ (Fig. 1(c)). It is observed that a linear fit is obtained in the high field region for all the temperatures. The values of dynamic dielectric constant, ε, extracted from the slope of the linear fit in the Schottky plots comes out to be in the range of 6.2 - 17 which is still significantly higher than the expected value of $\varepsilon \sim n^2 = 2.56$. Therefore, Schottky mechanism is not dominant conduction mechanism in the high field region.

The final mechanism is the space charge limited current (SCLC), in which conduction is controlled by discrete trap levels in the band gap of the dielectric. When the current density J dominated by SCLC the expected dependence on electric field is given by [21]

$$J = \frac{9}{8}\mu\varepsilon_0\varepsilon_r\theta\frac{E^2}{d} \tag{3}$$

where μ is the electron mobility, ε_r is the relative dielectric constant, ε_o is the permittivity of the free space, E is the applied electric field, d is the film thickness, and θ is the ratio of free to trapped charges. To evaluate the SCLC mechanism the I-V data is plotted as $ln\ J$ versus $ln\ E$ as shown in Fig. 2(a). It is observed that a linear fit is obtained in the high field region for all the temperatures. Again similar results were obtained for pure alumina films shown in Fig. 3(a). The slopes obtained from the $ln\ J$ versus $ln\ E$ plots in the high field region for SCLC mechanism for both the alumina/silicone nanolaminates and single layer alumina films come nearly about 2. Therefore, it is concluded that the SCLC mechanism is the dominant mechanism controlling the leakage current for field strength > 0.2 MV/cm for these high performance dielectrics. The SCLC mechanism is not consistent with the data at low field strength. This change in mechanism is attributed to the changes in the injected free carrier density, n_i. At low field region the value of n_i is lower compared to the intrinsic free carrier density (n_o) whereas the injected free carrier density (n_i) exceeds n_o in the high field region in these nanolaminates. As the injected free carrier density increases in the high field region the SCLC mechanism becomes dominant [20].

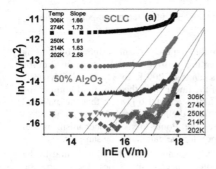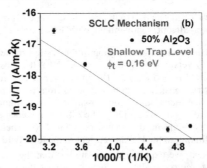

Figure 2. lnJ vs. lnE showing SCLC mechanism in the high field region (a) and ln(J/T) vs. 1000/T (b) of Alumina-silicone nanolaminates with 50% Al_2O_3. Instrumental error of $\pm$ 5% is shown as error bar in (b).

The SCLC mechanism was further examined through analysis of the trap energy levels. If the traps present in the dielectric film are shallow, θ can be given by [22]

$$\theta = \frac{N_C}{N_t}\exp\left[\frac{-(E_C-E_t)}{kT}\right] \tag{4}$$

where N_t is the trap density, N_C is the density of states in the conduction band, $(E_C - E_t)$ is the activation energy for the shallow electron traps, and T is the device temperature. Since N_C in θ in Eq. (4) is proportional to $T^{3/2}$ and the electron mobility μ in Eq. (3) is proportional to $T^{-1/2}$, therefore the temperature dependence of the current density J in the case of SCLC mechanism is given by [22]

$$J \propto T \exp\left[\frac{-(E_C - E_t)}{kT}\right] \tag{5}$$

To obtain the shallow electron trap level energy for the SCLC conduction process, we have plotted ln (J/T) versus $1000/T$ at a fixed electric field of 5.75×10^7 V/m for the 50% Al_2O_3 nanolaminates as shown in Fig. 2(b). The shallow trap level energy (E_t) estimated from the slope of the linear fits comes out to be 0.16 eV.

From figure 1(a), it is observed that in the low field region the leakage current data points are deviating from the linear fit. This concluded that some other conduction mechanism is contributing to the leakage current instead of SCLC mechanism in the low field regime.

As the leakage current density is temperature dependent (Fig.1(a)) also in the low field region, the direct tunneling mechanism is ruled out. At this low field regime, the possible leakage mechanism is Ohmic conduction which is carried by thermally excited electrons hopping from one state to the next. This Ohmic current is given by [21]

$$J \sim E \exp(-E_a/kT) \tag{6}$$

where E is the electric field, T is the device temperature, and E_a is the thermal activation energy of conduction electrons.

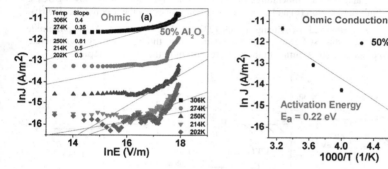

Figure 3. lnJ vs. ln E (a) and ln J vs. 1000/T (b) for Alumina-silicone nanolaminates with 50% Al_2O_3. Instrumental error of ± 5% is shown as error bar in (b).

To further confirm the dominance of Ohmic conduction mechanism, the I-V data is plotted as ln J versus ln E according to Eq. (6) and is shown in Fig. 3(a). It can be seen that the I-V data in the low field regime perfectly fits to a straight line with a slope ~ 1. Hence, Ohmic conduction is the dominant mechanism in the low electric field region. Therefore, one can conclude that the conduction process through alumina/silicone nanolaminates is Ohmic conduction in the low field region whereas SCLC mechanism is dominant in the high field region.

In the low field region where Ohmic conduction is dominant, the plot of ln J versus $1000/T$ gives a straight line at a fixed electric field as per Eq. (6) and the slope determines the thermal activation energy (E_a) for the Ohmic conduction mechanism. The activation energy was determined at a fixed electric field for alumina/silicone nanolaminate from the $ln\,J$ versus $1000/T$ plot as shown in Fiure 3(b) and it comes out to be 0.22 eV.

The temperature dependent I-V characteristics are also taken for pure single layer alumina films and leakage mechanisms in different field regions are studied. It is observed that the SCLC mechanism is dominant in the high field region (Fig. 4(a)) whereas Ohmic conduction contributes to the leakage current in the low field region (Fig. 4(b)).

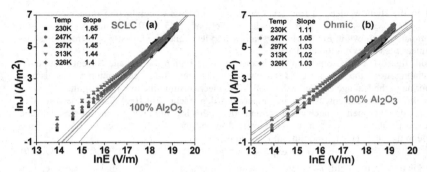

Figure 4. ln J vs. ln E at different temperatures for100% Al_2O_3 single layer films showing SCLC mechanism in the high field region (a) and Ohmic conduction process in the low field region (b).

A similar finding was reported by Zhang et al [23] in their study of Al_2O_3-polyimide nanocomposite dielectrics that SCLC mechanism is dominant in the high field region whereas Ohmic conduction process is contributing to the leakage current in the low field region.

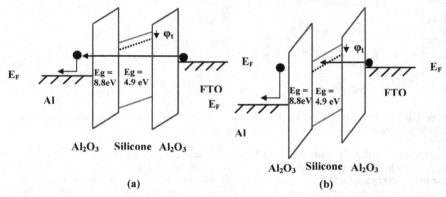

Figure 5. Energy band diagram showing Ohmic conduction in low field region (a) and SCLC conduction mechanism in high field region (b).

Figure 5 shows an energy band diagram to explain the various leakage current mechanisms responsible for the carrier transport in the nanolaminates in different field regions. In the low field region (Figure 5(a)) the electrons can directly tunnel where the shallow traps do not participate in the conduction process so that ohmic conduction occurs [24]. In the high field region (Figure 5(b)), more band bending occurs. Thus, the injected electrons from the bottom electrode (FTO) are trapped by the shallow trap level below the conduction band in silicone polymer dielectric [25] and then detrapped and goes towards the top electrode (Al) contributing to the SCLC current. Therefore, at high field region, SCLC mechanism occurs for the conduction process.

CONCLUSIONS

In conclusion, the mechanism of electron transport in alumina/silicone nanolaminates has been studied using temperature-dependent I-V measurements and analysis. After considering a number of alternatives it was found that the SCLC mechanism is the dominant conduction process at high field region whereas Ohmic conduction process is contributing to the leakage current in low field region in high performance nanolaminates with 50% Al_2O_3 content. Shallow trap level of energy 0.16 eV is responsible for the SCLC mechanism while activation energy of 0.22 eV for electrons is responsible for Ohmic conduction process in the nanolaminates. Similar types of conduction mechanism are also responsible for the leakage current in pure single layer Al_2O_3 films.

ACKNOWLEDGEMENTS

We gratefully acknowledge the National Science Foundation for the support of this work through award CMMI-0826323 and would like to thank the National Renewable Energy Laboratory, Golden, Colorado for using their electrical measurement facilities.

REFERENCES

1. R. P. Ortiz, A. Facchetti, and T. J. Marks, Chem. Rev. **110**, 205 (2010).
2. M. C. Choi, Y. Kim, and C. S. Ha, Prog. Polym. Sci. **33**, 581 (2008).
3. G. Dennler, C. Lungenschmied, H. Neugebauer, N. S. Sariciftci, M. Latrèche, G. Czeremuszkin, and M. R. Wertheimer, Thin Solid Films **511-512**, 349 (2006).
4. A. L. Deman, M. Erouel, D. Lallemand, M. Phaner-Goutorbe, P. Lang, and J. Tardy, J. Non-Cryst. Solids **354**, 1598 (2008).
5. D. K. Hwang, W. Choi, J.-M. Choi, K. Lee, J. H. Park, E. Kim, J. H. Kim and S. Im, J. Electrochem. Soc. **154**, H933 (2007).
6. D. K. Hwang, C. S. Kim, J. M. Choi, K. Lee, J. H. Park, E. Kim, H. K. Baik, J. H. Kim, and S. Im, Adv. Mater. **18**, 2299 (2006).
7. K. Kukli, J. Ihanus, M. Ritala, and M. Leskelä, J. Electrochem. Soc. **144**, 300 (1997).
8. H. Kattelus, M. Ylilammi, J. Saarilahti, J. Antson and S. Lindfors, Thin Solid Films **225**, 296 (1993).
9. P. C. Rowlette and C. A. Wolden, Thin Solid Films **518**, 3337 (2010).
10. J.-M. Choi, K. Lee, D. K. Hwang, J. H. Park, E. Kim, and S. Im, Electrochem. Solid-State Lett. **9**, G289 (2006).
11. Y. G. Seol, J. S. Park, N. T. Tien, N. E. Lee, D. K. Lee, S. C. Lee, Y. J. Kim, C. S. Lee, and H. Kim, J. Electrochem. Soc. **157**, H1046 (2010).

12. K. Choi, D. K. Hwang, K. Lee, J. H. Kim, and S. Im, Electrochem. Solid State Lett. **10**, H114 (2007).

13. S. K. Sahoo, D. Misra, D. C. Agrawal, Y. N. Mohapatra, S. B. Majumder, and R. S. Katiyar, J. Appl. Phys. **109**, 064108 (2011).

14. S. K. Sahoo, D. C. Agrawal, Y. N. Mohapatra, S. B. Majumder, and R. S. Katiyar, Appl. Phys. Lett. **85**, 5001 (2004).

15. E. L. Murphy and R. H. Good, Jr., Phys. Rev. **102**, 1464 (1956).

16. S. K. Sahoo and D. Misra, J. Appl. Phys. **110**, 084104 (2011).

17. S. -J. Ding, J. Xu, Y. Huang, Q. -Q. Sun, D. W. Zhang, and M. -F. Li, Appl. Phys. Lett. **93**, 092909 (2008).

18. R. P. Patel, D. Chiavetta, and C. A. Wolden, J. Vac. Sci. Technol. A **29**, 061508-1 (2011).

19. R. P. Patel and C. A. Wolden, J. Vac. Sci. Technol. A **29**, 021012 (2011).

20. S. K. Sahoo, R. P. Patel, and C. A. Wolden, Appl. Phys. Lett. 101, 142903 (2012).

21. S. M. Sze, *Physics of Semiconductor Devices*, 2nd ed. (Wiley-Interscience, 1981).

22. I. Y. -K. Chang and J. Y. -M. Lee, Appl. Phys. Lett. **93**, 223503 (2008).

23. P. Zhang, F. Chen, Y. Liu, and Q. Lei, Annual Report Conference on Electrical Insulation and Dielectric Phenomena, 260 (2007).

24. S. K. Sahoo and D. Misra, Appl. Phys. Lett. **100**, 232903 (2012).

25. H. Zhou, J. A. Dorman, Y. -C. Perng, J. P. Chang, and J. Liu, J. Appl. Phys. **111**, 064505 (2012).

Mater. Res. Soc. Symp. Proc. Vol. 1547 © 2013 Materials Research Society
DOI: 10.1557/opl.2013.542

Growth of silicon nanowires-based heterostructures and their plasmonic modeling

Yuan Li[1], Wenwu Shi[1], John C. Dykes[3], and Nitin Chopra[1,2]
[1]Metallurgical and Materials Engineering Department, Center for Materials for Information Technology (MINT), The University of Alabama, Tuscaloosa, AL 35487, U.S.A.
[2]Department of Biological Sciences, The University of Alabama, Tuscaloosa, AL 35487, U.S.A.
[3] REU, Department of Mathematics, The University of Alabama, Tuscaloosa, AL 35487, U.S.A
*Corresponding Author E mail: nchopra@eng.ua.edu, Tel: 205-348-4153, Fax: 205-348-2164

ABSTRACT

Complex nanoscale architectures based on gold nanoparticles (AuNPs) can result in spatially-resolved plasmonics. Herein, we demonstrate the growth of silicon nanowires (SiNWs), heterostructures of SiNWs decorated with AuNPs, and SiNWs decorated with graphene shells encapsulated gold nanoparticles (GNPs). The fabrication approach combined CVD growth of nanowires and graphene with direct nucleation of AuNPs. The plasmonic or optical properties of SiNWs and their complex heterostructures were simulated using discrete dipole approximation method. Extinction efficiency spectra peak for SiNW significantly red-shifted (from 512 nm to 597 nm or 674 nm) after decoration with AuNPs, irrespective of the incident wave vector. Finally, SiNW decorated with GNPs resulted in incident wave vector-dependent extinction efficiency peak. For this case, wave vector aligned with the nanowire axial direction showed a broad peak at ~535 nm. However, significant scattering and no peak was observed when aligned in radial direction of the SiNWs. Such spatially-resolved and tunable plasmonic or optical properties of nanoscale heterostructures hold strong potential for optical sensor and devices.

INTRODUCTION

Enhanced light-matter interactions for the molecules absorbed on noble metal nanostructures hold potential for sensitive optical sensors [1-3]. For example, anomalously high Raman signals can involve chemical and electromagnetic effects [23-5]. The latter effect relies on the oscillations of surface electrons (plasmon) when noble metal nanostructures are excited by incident light of a specific wavelength. Apart from wavelength of the incident illumination, morphology, and packing and geometrical arrangement of nanostructures as well as substrates are critical factors [6-10]. The generation of electromagnetic field (near field) under illumination can be theoretically calculated by solving 3-D Maxwell equations using discrete dipole approximation (DDA), finite difference time domain (FDTD), discontinuous Galerkin time domain (DGTD), and finite element method (FEM) [11]. Among these, discrete dipole scattering (DDSCAT) based on DDA is widely used due to its simplicity and ability for solving complex and irregular targets [12,13]. Of particular interest is a combination of experimental and theoretical approaches that could allow for discovery of unique nanoscale plasmonic architectures. Here, we report growth and plasmonic modeling of silicon nanowires (SiNWs), SiNWs decorated with gold nanoparticles (AuNPs) referred to as SiNWs-AuNPs heterostructures, and SiNWs decorated with graphene shells encapsulated AuNPs (SiNWs-GNP heterostructures).

EXPERIMENT

Materials and methods: N-type, (100) silicon (Si) wafers were purchased from IWS (Colfax, CA). Gold (III) chloride trihydrate (HAuCl$_4$·3H$_2$O, 99.9%) was purchased from Sigma-Aldrich (St. Louis, MO). Sodium borohydride (NaBH$_4$, powder, 98+%) was bought from Acros Organics (New Jersey, NJ). DI water (18.1 MΩ-cm) was obtained using a Barnstead International DI water system (E-pure D4641. Growth and annealing of the SiNWs was conducted inside a GSL-1100X Tube Furnace (MTI Corporation). H$_2$ (40% balanced with Ar), SiH$_4$ and Ar (all UHP grade) gas cylinders were purchased from Airgas South (Tuscaloosa, AL). Oxygen plasma treatment was performed in a Nordson March Jupiter III Reactive Ion Etcher (Concord, CA).

Growth of silicon nanowires: Si wafer substrate was oxidized with 3:1 H$_2$SO$_4$: H$_2$O$_2$ piranha solution for 30 min at 100 °C, and a 5 nm-thick gold film was sputtered on the substrate. SiNWs were grown in atmospheric pressure Chemical Vapor Deposition (CVD) process. The furnace was first heated to 850 °C in inert environment of 5% H$_2$/Ar. The temperature was kept for 10 min to produce the gold nano-droplet catalysts. Then the temperature was decreased to the growth temperature (625 °C). Approximately 2% SiH$_4$ in He, as Si source, was fed into the quartz tube with the carrier gas of 10% H$_2$ in Ar. The duration was 1 min for the growth of short nanowire and 5 min for the growth of long nanowires. Finally, the furnace was cooled down to room temperature in Ar environment.

Preparation of silicon nanowire-gold nanoparticles heterostructures: As-produced SiNWs were dispersed in ethanol by sonication. Approximately, 10 µL of sodium borohydride (NaBH$_4$, 0.12 M) was first added to 5 mL of the SiNW suspension. After 1 min of stirring, 100 µL HAuCl$_4$ (5×10^{-3} M) was added to the suspension and the reaction was carried on for 10 min. The resulting precipitate (SiNWs-AuNPs heterostructures) was washed with copious amounts of alcohol and re-dispersed in 1 ml ethanol. SiNWs-AuNPs heterostructures (10 µL) were then drop-casted on the oxidized Si wafer (1 cm × 1 cm) and dried in air. This sample was annealed at ~625 °C for 45 min to uniformly distribute AuNPs on SiNWs.

Growth of graphene shells encapsulating gold nanoparticles: As-prepared SiNWs-AuNPs heterostructures were treated in buffered oxide etch (BOE) solution for 10 s to etch the native oxide and then plasma oxidized for 15 min to result in surface gold oxide. Growth of graphene shells was conducted in a CVD process at 675 °C [14,15]. The carrier gas was 10% H$_2$/Ar. Xylene was used as the carbon source. The growth was continued for 1 h. This resulted in SiNWs-GNP heterostructures.

Discrete dipole approximation (DDA) calculations: Discrete Dipole Approximation (DDSCAT 7.2 software) was utilized to calculate the scattering and absorption of incident electromagnetic wave by different targets with arbitrary geometries and complex refractive index [12,13]. Different targets were constructed using Autodesk 3D max® software and dimensions were directly obtained from microscopic images. The normalized electric field distribution was calculated at the peak wavelength using DDSCAT 7.2 software and plotted using Mayavi2.

RESULTS AND DISCUSSION

Figure 1A and B shows microscopic images of SiNWs prepared from vapor-liquid-solid (VLS) method as evidenced by the presence of AuNP at the tip of the nanowires (Figure 1A inset). The diameter of SiNWs was observed to be ~110.6±40.4 nm and closely matched with AuNPs at the tips. These nanowires provided a unique substrate for further loading of AuNPs, where improved light-matter interaction was anticipated [7]. It was observed that average AuNP size and inter-particle spacing after direct nucleation process was ~10.4±3.6 nm and 6.2±2.3 nm, respectively. However, the SiNWs were contaminated with chemicals during the wet-chemical nucleation process. Thus, the heterostructures were annealed at ~625 °C for 45 min and resulted in average AuNP size of ~18.7±9.8 nm and inter-particle spacing of ~10.0±4.1 nm (Figure 1C). To provide robust surface passivation to these heterostructures, SiNWs-AuNPs heterostructures were further encapsulated in graphitic or carbon shells to result in SiNWs-GNP heterostructures (Figures 1D-F). For the growth of graphene shells on AuNPs, it has been demonstrated that surface oxidation of the latter is a critical step that facilitates the electron transfer reaction between surface gold oxide and incoming carbon feed, at CVD growth temperatures, to result in GNPs [14,15]. This is the reason for plasma oxidizing the SiNWs-AuNPs heterostructures prior to hydrocarbon CVD growth step. After the growth, it was observed that graphitic shells only encapsulated AuNPs and amorphous carbon was present on the exposed regions of SiNWs (Figure 1F). These graphene shells were observed to be ~2.3±0.5 nm thick with encapsulated AuNPs (diameter ~ 23.4±8.3 nm) and inter-particles spacing of ~8.3±2.9 nm, respectively. This is an interesting observation as it shows that due to the growth of graphene shells, surface migration of AuNPs was restricted at graphene shell growth temperatures. This is evidenced by diameters and inter-particle spacing for AuNPs, both of which remained nearly similar before and after graphene shells growth.

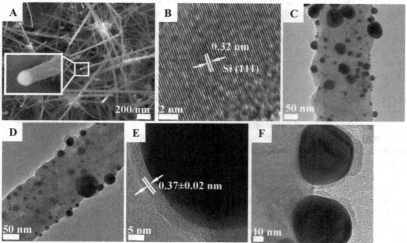

Figure 1. SEM and TEM images of (A and B) SiNWs, (C) SiNWs-AuNPs heterostructures, (D-F) SiNWs-GNP heterostructures.

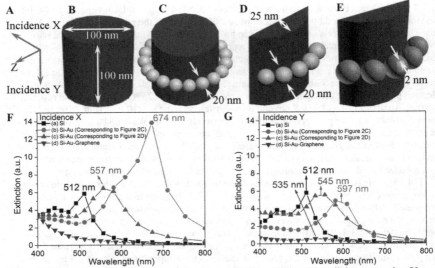

Figure 2. Schematics showing (A) two directions of incident lights (X: radial to nanowire, Y: axial to nanowire), (B) SiNW, (C) SiNW-AuNPs, (D) a section of SiNW-AuNPs, (E) a section of SiNW coated with GNPs (shell ~2 nm thickness), (F) and (G) comparison of extinction spectra for two directions for k vectors (X and Y) and peak locations.

Based on the calculation of diameters of SiNWs and AuNPs, we constructed four types of targets for calculation of extinction spectra (scattering and absorption) and mapped normalized electric field at the peak extinction wavelength. Since the targets are anisotropic, two incident directions (X: radial and Y: axial, Figure 2A) were considered. Figure 2B-E show four targets: (a) SiNW, (b) SiNW-AuNPs heterostructure, (c) a section of SiNWs-AuNPs heterostructure from (b), and (d) SiNWs-GNP heterostructure corresponding to section in (c) and with 2 nm thick graphene shells. Due to the much thinner graphene shells as compared to the diameters of AuNPs and SiNW, only sections of heterostructures were considered, where number of dipoles are less than a million and enough dipoles were allocated to 2 nm graphene shells [12]. This could significantly reduce the computation time and improve accuracy. As seen in Figure 2F and 2G, SiNW for X and Y directions showed extinction peaks at 512 nm. For SiNWs-AuNPs heterostructures, plasmonic peak red-shifted to ~674 nm for X direction and ~597 nm for Y direction. This is from the electromagnetic coupling between AuNPs and SiNW. Peak intensity also differed for different orientations of incoming wave vectors. Extinction peak intensity in X direction for SiNW decorated with nanoparticles is 2.3 times higher than intensity in Y direction. This indicates that electromagnetic field effects were stronger when incident wave propagated in the radial direction rather than axial direction. When only a section of heterostructure was considered (Figure 2D), peaks shifted toward lower wavelength (557 nm at X direction and 545 nm at Y direction). The shift of peaks and change of intensity indicated that geometry of SiNWs and distribution density of AuNPs is critical for plasmonic properties. However, SiNWs-GNP heterostructures (with 2 nm shell thickness), extinction spectra at both directions diminished sharply. Only a broad peak at ~535 nm was observed in the Y direction but scattering dominated in X. The reduction in peak intensity could be attributed to the thickness of graphene shells. Overall, it shows that optical properties after graphene shell encapsulation can be spatially resolved and depends on the direction of incident wave vector. It is also anticipated that to retain optical properties of encapsulated AuNPs, the shell thickness should be lower than 2 nm [14,15].

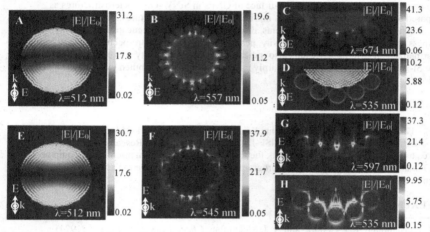

Figure 3. Distribution of normalized electric field in X direction for (A) SiNW, (B) SiNW-AuNPs, (C) a section of (B), (D) SiNW-GNP heterostructures and in Y direction for (E) SiNW, (F) SiNW-AuNPs, (G) a section of (F), (H) SiNW-GNP heterostructures.

Normalized electric field distributions were mapped using extinction results (Figure 3). Three major conclusions could be derived. (1) The field or plasmons are concentrated at the interface between Si and Au. This has also been observed by others and is mainly caused by the induction of image charge on substrate surface as well as multiple scattering and reflection by nanoparticles [16]. (2) Normalized electric field could be enhanced by decoration of AuNPs. However, the increment is dependent on the direction of incident wave vector (increase from 30.7 in X direction to 37.9 in Y direction). (3) Graphene shells (~2 nm) greatly depressed the intensity for both directions (75% drop in X direction and 72.5% decrease in Y direction) but enhanced the distribution of plasmons over larger surface area of GNPs as compared to only interfacial hot spots in case of SiNW-AuNP heterostructures.

CONCLUSIONS

CVD approach is used to grow SiNWs which were then utilized to nucleate AuNPs resulting in SiNWs-AuNPs heterostructures. These heterostructures were further surface plasma oxidized to allow for selective growth of graphitic shells encapsulating AuNPs while the exposed regions of SiNWs were encapsulated within an amorphous carbon. Such hybrid combinations based on SiNWs were further studied for their optical properties using DDA calculations. It was estimated that the bare SiNWs showed an extinction efficiency peak at ~512 nm, which was significantly red shifted to ~674 nm after decorating with the AuNPs. However, this activity is suppressed when they are encapsulated within graphene shells. At the same time, presence of graphene shells make the heterostructures spatially-resolved with a broad extinction efficiency peak at ~535 nm when the incident wave vector is in the nanowire axial direction and no peak in case of wave vector in nanowire radial direction. In addition, normalized electric field of plasmons were distributed over larger surface of GNPs in SiNW-GNP heterostructures as compared to being limited to interfaces in SiNW-AuNPs heterostructures.. Overall, incident light wavelength, morphologies of components, distribution density, and encapsulating shell thickness strongly influenced the intensity, peak location of extinction spectra, and distribution of normalized electric field. Results also indicate strong field intensity was mainly concentrated at the Si-Au interface and quenched sharply as well as broadly distributed by ~2 nm graphene shells.

ACKNOWLEDGMENTS

This work was funded by National Science Foundation (No. 0925445), the associated NSF-REU supplemental award, 2012 NSF-EPSCoR RII award, and Research Grant Committee awards to Dr. Chopra. The authors thank the CAF facility for electron microscopy equipment. J.C.D thanks UA's emerging scholars program (ESP) for his participation as undergraduate intern in this research.

REFERENCES

1. M. Fleischmann, P. J. Hendra, and A. J. McQuillan, *Chem. Phys. Lett.* **26**, 163 (1974).
2. A. Campion, and P. Kambhampati, *Chem. Soc. Rev.* **27**, 241 (1998).
3. Z. Q. Tian, B. Ren, and D. Y. Wu, *J. Phys. Chem. B* **106**, 9463 (2002).
4. S. K. Saikin, Y. Chu, D. Rappoport, K. B. Crozier, A. Aspuru-Guzik, *J. Phys. Chem. Lett.* **1**, 2740 (2010).
5. S. Timur, A. Vaskevich, I. Rubinstein, and G. Haran, *J. Am. Chem. Soc.* **131**, 14390 (2009).
6. M. Suzuki, Y. Niidome, Y. Kuwahara, N. Terasaki, K. Inoue, S. Yamada, *J. Phys. Chem. B* **108**, 11660 (2004).
7. N. J. Halas, S. Lal, S. Link, W. S. Chang, D. Natelson, J. H. Hafner, and P. Nordlander, *Adv. Mater.* **24**, 4774 (2012).
8. F. Cheng, A. Agarwal, K. D. Buddharaju, N. M. Khalid, S. M. Salim, E. Widjaja, M. V. Garland, N. Balasubramanian, and D. L. Kwong, *Biosensors Bioelectron.* **24**, 216 (2008).
9. M. Jin, V. Pully, C. Otto, A. van den Berg, and E. T. Carlen, *J. Phys. Chem. C* **114**, 21953 (2010).
10. N. A. A. Hatab, J. M. Oran, and M. J. Sepaniak, *ACS Nano* **2**, 377 (2008).
11. M. Karamehmedović, R. Schuh, V. Schmidt, T. Wriedt, C. Matyssek, W. Hergert, A. Stalmashonak, G. Seifert, and O. Stranik. *Opt. Express* **19**, 8939 (2011).
12. P. J. Flatau, and B. T. Draine, *Opt. Express*, **20**, 1247 (2012).
13. B. T. Draine, and P. J. Flatau, *J. Opt. Soc. Am. A* **11**, 1491 (1994).
14. N. Chopra, L. G. Bachas, and M. R. Knecht. *Chem. Mater.* **21**, 1176 (2009).
15. J. Wu, W. Shi, N. Chopra, *J. Phys. Chem. C* **116**, 12861 (2012).
16. P. A. Atanasov, N. N. Nedyalkov, T. Sakai, M. Obara. *Appl. Surf. Sci.* **254**, 794 (2005).

Mater. Res. Soc. Symp. Proc. Vol. 1547 © 2013 Materials Research Society
DOI: 10.1557/opl.2013.579

Luminescence enhancement of colloidal quantum dots by strain compensation

Y. Lu, Y.Q. Zhang, and X. A. Cao
Department of Computer Science and Electrical Engineering, West Virginia University,
Morgantown, WV 26506, U.S.A

ABSTRACT

We have investigated the effects of two different strain-relief bilayer shell structures on the luminescent properties of colloidal CdSe quantum dots (QDs). CdSe QDs with a strain-compensated ZnS/ZnCdS bilayer shell were synthesized using the successive ion layer adsorption and reaction technique and their crystallinity of was examined by X-ray diffraction. The QDs enjoyed the benefits of excellent exciton confinement by the ZnS intermediate shell and strain compensation by the ZnCdS outer shell. The resulting CdSe/ZnS/ZnCdS QDs exhibited 40% stronger photoluminescence and a smaller peak redshift upon shell growth compared to conventional CdSe/ZnCdS/ZnS core/shell/shell QDs with an intermediate lattice adaptor. CdSe/ZnS/ZnCdS QD light-emitting diodes (LEDs) had a luminance of 558 cd/m^2 at 20 mA/cm^2, 28% higher than that of CdSe/ZnCdS/ZnS QD-LEDs. The former also had better spectral purity. These results suggest that nanocrystal shells may be strain-engineered in a different way to achieve QDs of high crystalline and optical quality well suited for full-color display applications.

INTRODUCTION

Colloidal nanocrystal quantum dots (QDs) synthesized by low-cost solution processes have many attributes which make them suitable for optoelectronic applications [1-3]. However, the performance of QD-based optoelectronic devices, including light-emitting diodes (LEDs), and solar cells, is lagging significantly behind that of devices based on conventional bulk materials [1]. Due to the high surface-to-volume ratio inherent to nanocrystals, the device performance is largely determined by the surface properties of QDs. Surface defects trap carriers and enhance nonradiative recombination [3,4]. Organic ligands introduced for dispersion during colloidal synthesis may protect the surfaces of QDs in solutions, but they cannot provide full passivation of the QDs in device structures as many ligands may detach from the QD surfaces during device processing [4]. A common strategy is to better passivate the QD surface with a thin shell of a wider band gap semiconductor, forming core/shell (CS) nanoparticles [3]. The shell reduces the number of surface dangling bonds and physically separates the optically active core from its surrounding medium. Consequently, core/shell QDs have much improved photoluminescence (PL) quantum yield (QY) and stability against photo-oxidation as compared to QD cores.

Among many CS QDs, CdSe/ZnS CS QDs have been a subject of extensive research [3, 5-10]. ZnS possesses a much larger band gap than CdSe and forms a Type-I heterojunction with CdSe [3]. Even though ZnS provides effective electronic passivation of CdSe core QDs, it is not an ideal shell material because its lattice constant is ~12% smaller than that of CdSe. The large lattice mismatch prevents the epitaxial growth of a thick ZnS shell [7]. Tensile strain in the ZnS shell increases as the shell grows thicker, and eventually relaxes through misfit defect formation, degrading the optical properties of the CS QDs [6,8]. To mitigate the problem, a thin layer of a semiconductor whose lattice constant falls between those of CdSe and ZnS, such as CdS, ZnSe, or ZnCdS, is often inserted between the CdSe core and ZnS shell, forming a core/shell/shell (CSS) structure [7-10]. The intermediate shell acts as a lattice adapter and a strain-reducing layer, giving rise to a significant improvement of the structural and optical properties of the QDs.

In this work, we propose an alternative strategy to reduce strain in CdSe/ZnS QDs and prove its effectiveness in defect reduction and luminescence enhancement. A layer of a material with a lattice constant smaller than that of ZnS, i.e. ZnCdS, is grown outside the ZnS shell as a strain-compensating layer. The resulting CSS QDs with a strain-compensated shell are expected to have advantages over conventional CSS QDs since the wide bandgap ZnS shell directly covers the core, and provides better passivation and carrier confinement than a strain-reducing intermediate shell in conventional CSS QDs, which typically has a smaller band gap than ZnS. To validate this new strategy, two types of CSS QDs, CdSe/ZnCdS/ZnS QDs and CdSe/ZnS/ZnCdS QDs, denoted as CSS I QDs and CSS II QDs, respectively, were synthesized and characterized. It has been found that CSS II QDs have superior optical properties, and the corresponding QD-LEDs have a higher luminescence efficiency and better spectral purity.

EXPERIMENT

CdSe QDs capped with octadecylamine ligands were synthesized using CdO and Se as precursors by the conventional hot-injection method [1]. The QDs were precipitated by addition of methanol, separated, and centrifuged to remove side products and unreacted precursors. To synthesize CdSe/ZnS and CdSe/Zn$_{0.5}$Cd$_{0.5}$S CS QDs, up to 6 monolayers (MLs) of ZnS or Zn$_{0.5}$Cd$_{0.5}$S were grown on the CdSe cores in a hexane solution at 240 °C using the successive ion layer adsorption and reaction (SILAR) method [11], To grow CSS I and CSS II QDs, up to 3 MLs of Zn$_{0.5}$Cd$_{0.5}$S and ZnS outer shells were grown on the CdSe/ZnS and CdSe/Zn$_{0.5}$Cd$_{0.5}$S CS QDs, respectively, by SILAR. During all these shell growth, a 1 mml sample was extracted after the growth of each ML and washed by hexane for optical characterization. The absorption and PL spectra of the QDs dissolved in hexane were measured using a Hitachi U-3900H UV-VIS spectrophotometer and a Hitachi F-7000 fluorescence spectrophotometer, respectively. The PL QYs were estimated by comparing the QD PL intensity with that of standard Rhodamine 6G dye solutions with the same optical density at the same excitation wavelength.

The schematic cross-sections and energy band diagrams of the CSS QDs are shown in Fig. 1. As seen, stronger quantum confinement of carriers is expected in CSS II QDs. The crystallinity of the synthesized CSS QDs was confirmed by powder X-ray diffraction (XRD). Three distinct diffraction peaks were observed at 2θ values of 25.8°, 43.3°, and 51.2°, corresponding to the (111), (220), and (311) crystalline planes of cubic CdSe, respectively.

The QD-LED fabrication included spin-coating of polymer and QD layers on ITO/glass

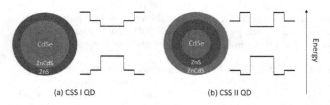

(a) CSS I QD   (b) CSS II QD

Figure 1. Schematic cross-sections and energy band diagrams of two different CSS QDs.

substrates, followed by evaporation deposition of organic and metal films. The device structure consisted of a 40 nm poly(3,4-ethylenedioxythiophene) poly(styrenesulfonate) (PEDOT:PSS) hole injection layer, a 40 nm poly-(N,N'-bis(4-butylphenyl)-N,N'-bis(phenyl)benzidine) (poly-TPD) hole transport layer, a CSS QD emitting layer (EML), and a 40 nm 2,2',2"-(1,3,5-benzenetriyl)-tris(L-phenyl-l-H-benzimidazole) (TPBi) electron transport layer. LiF/Al (0.5/120 nm) was deposited on top of the TPBi layer as the cathode, whose overlap with the ITO anode defines an emitting area of 0.1 cm^2. The QDs, being CdSe/2 ML $Zn_{0.5}Cd_{0.5}S$/2 ML ZnS CSS I QDs or CdSe/2 ML ZnS /2 ML $Zn_{0.5}Cd_{0.5}S$ CSS II QDs, were deposited by spin coating from 2 mg/ml QD/hexane solutions, resulting in a QD coverage of ~1.2 MLs. The QD-LEDs were encapsulated with glass lids in N_2 and characterized in air at room temperature. To calculate the external quantum efficiency (EQE), the QD-LEDs were placed directly on the surface of a calibrated silicon photodetector and all emitted photons from the glass side were captured.

RESULTS AND DISCUSSION

Figure 2 shows the evolution of the absorption spectra of the CS QDs (solid lines) and CSS QDs (dashed lines) during the shell growth. The CdSe core QDs have a first excitonic absorption peak at 521 nm. The full width at the half maximum of the peak is ~30 nm, suggesting a narrow size distribution. As the shell growth proceeds, the peak exhibits redshift and broadening due to the penetration of the electron and/or hole wave functions into the shell layers [3]. The redshift is more pronounced in the CdSe/ZnCdSe CS QDs than in the CdSe/ZnS CS QDs. The peak shifts by as much as ~53 nm in the CdSe/ZnCdSe (3 MLs) QDs, as compared to ~22 nm in CdSe/ZnS (3MLs) QDs. This is because, as seen from the schematic band structures in Fig. 1, the ZnCdS shell cannot provide potential barriers large enough to prevent the leakage of excitons into the shell [3], whereas the ZnS shell, whose bulk band gap is 0.6 eV larger, can confine carriers inside the cores more efficiently. However, the CdSe/ZnS CS QDs show more significant peak broadening, especially after the growth of 2-3 MLs of ZnS. This can be attributed to a larger lattice mismatch between CdSe and ZnS (as compared to CdSe/ZnCdSe), which may cause geometry irregularity or size variations [7]. As expected, growing a ZnS outer shell around the CdSe/3 ML ZnCdS CS QDs results in minimal further red shift, whereas growing a ZnCdS outer shell around the CdSe/3 ML ZnS CS QDs causes a small additional redshift (~11 nm). Interestingly, the former display significant peak broadening, indicative of degraded QD quality, whereas the latter show reduced peak broadening, evidencing the effect of strain compensation by the ZnCdS outer shell. Overall, the CSS II QDs exhibit more controllable peak wavelength and better quality than the CSS I QDs, due to more effective carrier confinement by the ZnS intermediate shell and more effective strain reduction by the compensating outer shell.

The evolution of the PL QY of the CSS QDs in solutions is shown in Fig. 3. The typical PL spectra are shown in the inset. The QY of the CdSe core QDs is 7.5%. The overgrowth of both ZnS and ZnCdS passivating shells leads to a dramatic improvement in the QY. It has been found that the QY of both CdSe/$Zn_{0.5}Cd_{0.5}S$ CS QDs and CdSe/ZnS CS QDs attains a maximum value of ~28% after the growth of 3 MLs of the shell material, and then decreases as the shell grows thicker. This behavior has been observed in many other CS QDs with a significant lattice mismatch between the core and shell [3]. The initial shell growth is coherent and strain builds up in the shell as it grows thicker. When a critical thickness is reached, strain is relieved through the formation of misfit dislocations at the interface, which act as nonradiative recombination sites,

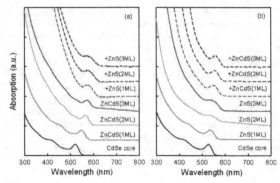

Figure 2. Evolution of the absorption spectra of (a) CdSe/ZnCdS/ZnS CSS I QDs and (b) CdSe/ZnS/ZnCdS CSS II QDs during the shell growth. The solid lines are spectra of CS QDs.

deteriorating the luminescent properties of the QDs [6-8]. As seen in Fig. 3, further growth of a ZnS outer shell around the CdSe/Zn$_{0.5}$Cd$_{0.5}$S(3MLs) CS QDs and a Zn$_{0.5}$Cd$_{0.5}$S outer shell around the CdSe/ZnS(3MLs) CS QDs leads to a further improvement of the PL QY. The CdSe/3 ML ZnCdS/ZnS QDs (CSS I QDs) acquire the highest QY of 36% after adding one ML of ZnS. A thicker ZnS outer shell, however, degrades the PL QYs. The PL QY of the CdSe/3 ML ZnS/ZnCdS QDs (CSS II QDs) achieves a more remarkable improvement and peaks at 48% with a 2 ML ZnCdS outer shell. Zn$_{0.5}$Cd$_{0.5}$S has a smaller lattice mismatch (~8%) with CdSe compared to ZnS. In the CSS I structure, it functions as a lattice adaptor and strain-reducing layer [10], whereas in the CSS II structure, it partially compensates the strain in the ZnS shell, giving rise to a strain-compensated bilayer shell structure. So strain relief can be achieved in both CSS QDs, but by different mechanisms. The results in Fig. 3 suggest that the strain compensation strategy is more effective in improving the quality of the QDs. In this special case, the CSS II QDs take the benefits of excellent exciton confinement by a ZnS intermediate shell and strain compensation by a ZnCdS outer shell. As seen in Fig. 2, another advantage of the CSS II structure is a much smaller peak redshift resulting from the shell growth.

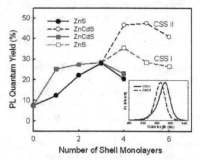

Figure 3. Evolution of the PL QY of CdSe CS and CSS QDs during the shell growth. The inset shows the PL spectra of two types of CSS QDs with a normalized peak intensity.

To investigate electroluminescence (EL) from the CSS QDs, LEDs with an EML comprising ~1.2 MLs of CdSe/2 ML ZnCdS/2 ML ZnS (CSS I) or CdSe/2 ML ZnS/2 ML ZnCdS (CSS II) QDs were fabricated and characterized. Figure 4 shows their EL spectra recorded at 0.2-20 mA/cm^2. In both LEDs, the EL originates from QDs only and no parasitic emission from the organic layers is observed. While the CSS II QD-LED emits purely yellow light with a peak at 583 nm, the EL spectrum of the CSS I QD-LED comprises weak broad emission in the blue-green wavelength region on the high-energy side of the main 595 nm peak. Such side emission, which appears to be weaker in the low injection regime, is absent in the PL of purified CSS I QDs. We believe that it arises from exciton recombination in the ZnCdS intermediate shell. The discrepancy between the EL and PL may be explained as follows. The optical excitation intensity of PL by a broadband lamp through a monochromator is low, so only a small number of excitons are generated, which are mainly confined and recombine in the CdSe core. However, under current injection, a much higher density of excitons may be created in the core through direct injection and energy transfer [5], which cannot be well confined by the ZnCdS intermediate shell. A small portion of the excitons may overflow into ZnCdS and undergo radiative decay. The broad band emission suggests that the ZnCdS shell may consist of Zn-rich and Cd-rich phases. Such compositional inhomogeneity proves that it is difficult to form a completely random ZnCdS alloy with a uniform composition due a large lattice mismatch between ZnS and CdS.

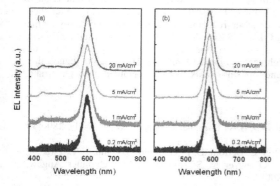

Figure 4. EL spectra of (a) CSS I and (b) CSS II QD-LEDs at increasing current density. The EL peak intensities are normalized, and the spectra are shifted in the y-direction for clarity.

Figure 5 shows the luminance-current-voltage (L-I-V) characteristics of the CSS I and CSS II QD-LEDs. Both devices turn on at ~4.2 V. Their operational voltages at 20 mA/cm^2 are 8.5 V and 7.7 V, respectively. The small difference may arise from different energy barriers for charge injection into these two types of CSS QDs due to their different shell structures. The CSS II QD-LED is brighter than the CSS I QD-LED. At 20 mA/cm^2, the luminance and current efficiency of the former are 558 cd/m^2 and 2.8 cd/A, respectively, which are 28% higher than those of the CSS I QD-LED. This EL improvement is consistent with the fact that the PL QY of the CSS II QDs was found to be ~40% higher than that of CSS I QDs. On the other hand, the EL efficiencies of the QD-LEDs are much lower than the PL efficiencies of the respective QDs measured in solutions. This can mainly be attributed to two factors [5]: (i) inefficient exciton

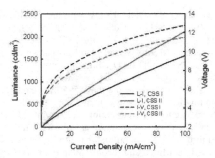

Figure 5. L-I-V characteristics of CSS I and CSS II QD-LEDs.

generation in the QD-LEDs due to poor charge injection into the QDs, and (ii) detachment of ligands from the QD surfaces during device processing, leading to degraded QD quality.

CONCLUSIONS

In summary, we have investigated the effects of two different strain-relief bilayer shell structures on the luminescence of colloidal CdSe QDs. A ZnCdS strain-compensating outer shell has proven to be more effective than a ZnCdS strain-reducing intermediate shell in improving the PL QY of CdSe/ZnS QDs. The resulting CdSe/ZnS/ZnCdS CSS QDs were 40% more efficient than the CdSe/ZnCdS/ZnS QDs, and have a more controllable peak wavelength. LEDs based on the former QDs exhibited 28% higher brightness and purer EL spectra. These results suggest that nanocrystal QDs of high crystalline and optical quality can be obtained through strain-engineering in their nanoshell structures.

REFERENCES

1. D. V. Talapin, J. S. Lee, M. V. Kovalenko, E. V. Shevchenko, Chem. Rev. **110** (1), 389-458 (2010).
2. A. L. Rogach, N. Gaponik, J. M. Lupton, C. Bertoni, D. E. Gallardo, S. Dunn, N. L. Pira, M. Paderi, P. Repetto, S. G. Romanov, C. O'Dwyer, M. S. Torres, and A. Eychmuller, Angew. Chem., Int. Ed. **47** (35), 6538-6549 (2008).
3. P. Reiss, M. Protie`re, and L. Li, Small **5** (2), 154–168 (2009).
4. A. Mews, Z. Phys. Chem. **221**, 295-306 (2007).
5. Y. Q. Zhang, X. A. Cao, Appl. Phys. Lett. **97**, 253115 (2010).
6. B. O. Dabbousi, J. Rodriguez-Viejo, F. V. Mikulec, J. R. Heine, H. Mattoussi, R. Ober, K. F. Jensen, M. Bawendi, J. Phys. Chem. B **101** (46), 9463-9475 (1997).
7. P. Reiss, S. Carayon, J. Bleuse, A. Pron, Synthetic Metals **139** (3), 649-652 (2003).
8. D. V. Talapin, I. Mekis, S. Goltzinger, A. Kornowski, O. Benson, H. Weller, J. Phys. Chem. B **108** (49), 18826-18831 (2004).
9. J. McBride, J. Treadway, L. C. Feldman, S. J. Pennycook, S. J. Rosenthal, Nano Lett. **6** (7), 1496-1501 (2006).
10. R. Xie, U. Kolb, J. Li, T. Basche, A. Mews, J. Am. Chem. Soc. **127** (20), 7480-7488 (2005).
11. J. J. Li, Y. A. Wang, W. Z. Guo, J. C. Keay, T. D. Mishima, M. B. Johnson, and X. G. Peng, J. Am. Chem. Soc. **125** (41), 12567-12575 (2003).

Mater. Res. Soc. Symp. Proc. Vol. 1547 © 2013 Materials Research Society
DOI: 10.1557/opl.2013.666

Enhanced visible-light absorption of mesoporous TiO$_2$ by co-doping with transition-metal/nitrogen ions

J. E. Mathis,[1,2] Z. Bi,[2] C. A. Bridges,[2] M. K. Kidder,[2] and M. P. Paranthaman[2]
[1]Physical Sciences Dept., Embry-Riddle Aeronautical University, Daytona Beach, FL 32114
[2]Chemical Sciences Division, Oak Ridge National Laboratory, Oak Ridge, TN 37831

ABSTRACT

Titanium (IV) oxide, TiO$_2$, has been the object of intense scrutiny for energy applications. TiO$_2$ is inexpensive, non-toxic, and has excellent corrosion resistance when exposed to electrolytes. A major drawback preventing the widespread use TiO$_2$ for photolysis is its relatively large band gap of ~3eV. Only light with wavelengths shorter than 400 nm, which is in the ultraviolet portion of the spectrum, has sufficient energy to be absorbed. Less than 14 percent of the solar irradiation reaching the earth's surface has energy exceeding this band gap. Adding dopants such as transition metals has long been used to reduce the gap and increase photocatalytic activity by accessing the visible part of the solar spectrum. The degree to which the band gap is reduced using transition metals depends in part on the overlap of the d-orbitals of the transition metals with the oxygen p-orbitals. Therefore, doping with anions such as nitrogen to modify the cation-anion orbital overlap is another approach to reduce the gap. Recent studies suggest that using a combination of transition metals and nitrogen as dopants is more effective at introducing intermediate states within the band gap, effectively narrowing it. Here we report the synthesis of mesoporous TiO$_2$ spheres, co-doped with transition metals and nitrogen that exhibit a nearly flat absorbance response across the visible spectrum extending into the near infrared.

INTRODUCTION

Modification of the composition and morphology of titanium (IV) oxide, TiO$_2$, continues to be an active area of research for applications in energy production and storage. In the area of energy production, TiO$_2$ can be used in photovoltaic or photocatalysis applications, particularly in splitting water into hydrogen and oxygen. TiO$_2$ is being investigated for use in energy storage as an electrode in lithium-ion batteries, to replace the graphite anodes presently used. A major obstacle in using TiO$_2$ for photolysis is its relatively large band gap of ~3eV. To excite an electron from the valence band to the conduction band requires the energy of the light striking TiO$_2$ to exceed this value. Only light with wavelengths shorter than 400 nm, which is in the ultraviolet portion of the spectrum, fulfills this requirement. Other intrinsic properties that hinder TiO$_2$'s employment as an anode for electrochemical cells include poor ionic conductivity and high electrical resistance.

Adding cationic dopants such as transition metals to TiO$_2$ has long been used to reduce this gap and increase photocatalytic activity [1-6]. In particular, Li, *et al.*, found that, chromium-, nitrogen-codoped TiO$_2$ (denoted as (Cr, N)TiO$_2$) exhibited promising photocatalytic properties[4]. An important milestone was the recognition that anions such as nitrogen may also act as dopants to reduce the gap [7]. A computational model indicates that the transition metals alter the density of electronic states immediately below the valence band edge, while the *p*-orbitals of nitrogen populate the gap just below the conduction band edge[8]. Using a combination of transition metals and nitrogen as dopants produces n-p co-doping of TiO_2. It has been postulated that this stabilizes the dopants because the electrostatic attraction between them overcomes the thermodynamic instability and enhances the kinetic solubility of the dopants [9]. The co-doping can be either "compensated," where the cation and anion charges balance, or "non-compensated" structures [10]. The optical absorption coefficient of a substance is a measure of the band gap in the region of interest. In this case, a large absorbance coefficient in the visible to near infrared region is highly desirable. These results spurred the present, systematic study of the effect of co-doping TiO_2 with comparing the optical characteristics of co-doped, mesoporous TiO_2 to that of theory.

EXPERIMENTAL PROCEDURE

The hydrothermal method [11, 12] was used to produce mesoporous TiO_2 microparticles that were doped with both a transition metal and nitrogen. The hydrothermal method uses high pressure and an elevated temperature to effect the formation of crystalline material. The morphology of these particles was controlled by optimizing experimental conditions including temperature, reaction time pH and concentration. Chromium, manganese, iron, cobalt, nickel, copper, and zinc comprised the group of transition metals tested. An amount of 20% $TiCl_3$ in 3% HCl corresponding to 0.0975 mmol $TiCl_3$ was added to 0.0025 mmol of the transition metal, 25 mmol urea, and 83 mmol de-ionized water. This was dissolved in 8.5 mL ethanol. The solution was stirred overnight. The solution was then transferred to a Teflon™-lined autoclave, which was heated to nearly 200°C. After this heat treatment, the samples were filtered, rinsed with de-ionized water, and dried. Nitrogen doping was accomplished by annealing the metal-doped TiO_2 samples in a 525°C furnace under an ammonia gas flow.

The samples were imaged using a Hitachi S-4800 scanning electron microscope. X-ray diffraction data were used to identify the crystalline phases present in the samples, and data were collected using a D5005 diffractometer and Cu K_α radiation. Determination of mesoporosity, surface area, and average pore size was accomplished using Brunauer, Emmett, Teller (BET) analysis. The samples' absorbance of light in the ultraviolet-visible (UV-vis) region was measured using a Varian Cary 5000 diffuse-reflectance spectrometer.

RESULTS AND DISCUSSION

Scanning electron microscopy confirmed that micron-sized, spherical particles were produced as shown in Figure 1. As is typical with hydrothermally produced material, the sizes were not uniform. The sizes ranged from 1 μm to approximately 3 μm.

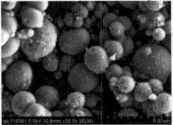

Figure 1 SEM image of chromium-doped TiO_2 particles produced by the hydrothermal method.

The BET gas adsorption profile measurements were performed on four of the samples. The profiles in Figure 2a show the presence of mesopores from the nitrogen gas adsorption/desorption patterns that are characteristic of a type IV isotherm and hysteresis typical of cylindrical pores. The analyses gave an average surface area of 81 m^2/g and an average pore diameter of 7 nm. All samples were confirmed to contain only the crystalline anatase phase of TiO_2 by powder X-ray diffraction analysis as shown in Figure 2b. As mentioned earlier, the anatase phase is preferable for energy applications, owing to its more open architecture.

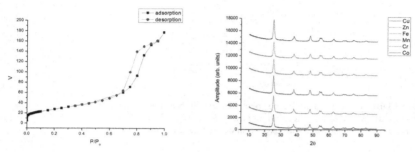

Figure 2 (a) BET isotherm plot of (Co, N) TiO_2 indicating a characteristic mesoporous structure. (b) X-ray diffraction plot of the samples displaying only the anatase phase.

UV-vis diffuse reflectance data reveal significant differences in the absorption spectrum as a function of doping. Without any dopants, the absorbance spectrum of native TiO_2 displays the typical TiO_2 response, with a steep decrease in absorbance starting near 400 nm. After introducing the metal dopants, the absorbance increased for all samples. The manganese-doped TiO_2 exhibited the highest level of absorbance from 500 nm to 600 nm, followed by chromium. When the metal-doped samples were then doped with nitrogen, the results were significantly modified. Nitrogen doping greatly increased the visible-region absorbance for all the samples and was dramatically larger for the (Co, N)TiO_2 sample as shown in Figure 3. As can be seen, the cobalt-doped TiO_2 absorbance from 500 nm to 600 nm surpassed all the others by a large margin. The (Fe, N)TiO_2 sample followed with the second-highest absorbance, and the (Zn, N)TiO_2 had the poorest absorbance of all.

These results can be compared with the expectations from a recent theoretical study [8] of transition metal/nitrogen co-doping, which predicted that manganese would have the largest absorption coefficient at 550 nm, the midpoint of the visible region of the spectrum, followed by copper and then cobalt. The overall order of strongest absorbance would be headed by Mn, followed by Cu, Co, Zn, Fe. Cr and Ni would have nearly equal absorbance values at 550 nm and the lowest absorbance. In contrast, we observed the largest absorbance with (Co,N) doping, and that absorbance for (Fe,N) doping was significantly higher than that of (Zn, N) doping. The different trend in absorbance with doping from the theoretical expectation is being further investigated, and may relate to differences in the level of metal and nitrogen doping that can be achieved through this synthetic approach.

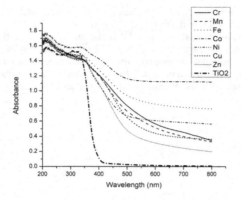

Figure 3 Diffuse UV-vis absorbance spectra of (metal, N) TiO_2 showing the dramatic variation in absorbance with the transition metals. The legend shows which metal was the dopant, with exception of the line showing undoped TiO_2.

CONCLUSIONS

This study demonstrated the capability to dramatically modify the UV-vis absorbance of TiO_2 through co-doping, while retaining the spherical mesoporous morphology of undoped, mesoporous TiO_2. Doping with nitrogen anions increases the absorbance in every case. The transition metals ions showing the greatest effect on the absorbance, cobalt and iron, differs from the expected trend of theoretical predictions. The large and flat absorbance over the visible spectrum bodes well for photovoltaic or photocatalytic applications.

ACKNOWLEDGEMENT

Materials synthesis work was sponsored by the U.S. Department of Energy, Basic Energy Sciences, Materials Sciences and Engineering Division. Characterization work was supported by Oak Ridge National Laboratory's CNMS, and SHaRE User Facility, which are sponsored by the Scientific User Facility Division, Office of Basic Energy Sciences, U. S. Department of Energy. JM is supported by ORISE through U.S. Department of Energy-Visiting Faculty Program (VFP). MKK acknowledges the support of the Division of Chemical Sciences, Geosciences, and Biosciences, Office of Basic Energy Sciences, U.S. Department of Energy. Thanks due to C. N. Sun, and G. M. Veith for assistance with BET measurements.

REFERENCES

[1] A. K. Ghosh and H. P. Muruska, *J. Electrochem. Soc.*, vol. 128 , pp. 1516-1522, 1977.
[2] P. Salvador, *Sol. Energy Mat.*, vol. 2 , pp. 413-421, 1980.

[3]   H. Yamashita, M. Harada, J. Misaka, M. Takeuchi, B. Neppolian and M. Anpo, *Catal. Today,* vol. 84 , pp. 191-196, 2003.

[4]   C. Wang, D. W. Bahnemann and J. K. Dohrman, *Chem. Comm.,* vol. 16 , pp. 1539-1540, 2000.

[5]   H. Yamashita, Y. Ichihashi, M. Takeuchi, S. Kishiguchi and M. Anpo, *J. Synchro, Rad.,* vol. 6 , pp. 451-452, 1999.

[6]   K. E. Karakitsou and X. E. Verykios, *J. Phys. Chem.,* vol. 97 , pp. 1184-1189, 1993.

[7]   R. Asahi, T. Morikawa, T. Ohwaki, K. Aoki and Y. Taga, *Science,* vol. 293, pp. 269-271, 2001.

[8]   R. Zhang, Q. Wang, J. Liang, Q. Li, J. Dai and W. Li, *Physica B,* vol. 407 , pp. 2709-2715, 2012.

[9]   Y. Li, W. Wang, X. Qiu, L. Song, H. M. M. III, M. P. Paranthaman, G. Eres, Z. Zhang and B. Gu, *App. Cat B: Env,* vol. 110, pp. 148-153, 2011.

[10] W. Zhu, X. Qiu, V. Iancu, X. Q. Chen, H. Pan, W. Wang, N. M. Dimitrijevic, T. Rajh, H. M. M. III, M. P. Paranthaman, G. M. Stocks, H. H. Weitering, B. Gu, G. Eres and Z. Zhang, *Phys. Rev. Lett.,* vol. 103, p. 226401, 2009.

[11] S. Yoon, A. Manthiram, J. Phys. Chem. C, vol. 115, p 9410 2011.

[12] S. Yoon, C.A. Bridges, R.R. Unocic, M.P. Paranthaman, J. Mater. Sci. in press, 2013.

Nanomaterials and Nanocomposites

Mater. Res. Soc. Symp. Proc. Vol. 1547 © 2013 Materials Research Society
DOI: 10.1557/opl.2013.783

Effects of Amines on Chemical and Physical Behaviors of Viscous Precursor Sols to Indium Gallium Zinc Oxide

Nobuko Fukuda, Shintaro Ogura, Ken-ichi Nomura, and Hirobumi Ushijima
Flexible Electronics Research Center (FLEC), National Institute of Advanced Industrial Science and Technology (AIST), 1-1-1 Higashi, Tsukuba 305-8565, Japan.

ABSTRACT

We synthesized viscous precursors to indium gallium zinc oxide (IGZO) using three kinds of alcoholamines, ethanolamine (EA), diethanolamine (DEA), and triethanolamine (TEA), by a simple process. The viscous precursors are obtained just by vigorous stirring of alcoholamine and urea in an aqueous solution containing the metal nitrates during heating at 150-160 °C. The precursor containing EA (EA-precursor) is a pale-orange suspension containing aggregates of the metal hydroxides and shows pseudoplastic flow. The precursors containing DEA (DEA-precursor) and TEA (TEA-precursor) are transparent pale-yellow and dark-orange sols, respectively. They give Newtonian flow in the lower shear rate and pseudoplastic flow in the higher shear rate. Higher concentration of metal salts leads to higher viscosity of the precursors. According to thermogravimetry-differential thermal analysis (TG-DTA) for the EA- and DEA-precursors, evaporation of alcoholamine occurs at around each boiling point and subsequently formation of metal oxides occur at around 300 °C. In the case of the TEA-precursor, formation of metal oxides occurs before pyrolysis of TEA attributed to the higher boiling point of TEA. The thin IGZO film, which is prepared by spin-coating of the diluted DEA-precursor and subsequent sintering at 450 °C for 30 min, shows 0.02 cm$^2 \cdot$V^{-1}s^{-1} of the mobility and 10^{-5} of the on/off ratio. The highly viscous DEA-precursor containing high concentration of metal ions allows patterning in an area of 100 cm^2 onto a surface of a silicon wafer with screen printing.

INTRODUCTION

Printing technology has great potentials for low-cost manufacturing of electronic and photoelectronic devices. The advantages of the printing technology include reduction of the number of manufacturing processes such as mask-less patterning, energy conservation, and suitability for manufacturing of flexible devices. Recently various printing processes and printable materials as inks have been widely developed for printed devices [1-3]. The printing methods are deeply related to chemical and physical properties of the inks. Viscosity is one of the key factors for determining the printing method. For example, appropriate viscosity for inkjet printing is lower than that for screen and offset printings. The most important points in development of printable materials are design and synthesis tailored to the desired printing method.

Indium-gallium-zinc oxide (IGZO) is an attractive material as a transparent metal oxide semiconductor [4]. Amorphous IGZO films prepared by sol-gel process and sintering show high performance as an n-typed semiconductor of a thin film transistor (TFT) [5]. Typical IGZO precursors are low-viscous sols with approximately 10 few mPa·s at most and spin-coating and inkjet printing are available. However, mass-printing methods such as screen and offset printings require higher-viscous precursors with more than 50 Pa·s at least for formation of micro-patterns. As compared with metal electrode materials for mass-printing methods, development of oxide semiconductor materials is behind except materials for solar photovoltaic cells such as zinc oxide and titanium dioxide [6]. Considering that the mass-printing methods lead to low-cost manufacturing, highly viscous materials should be further researched and developed.

In this work, we synthesized viscous IGZO precursors by a simple method for studying rheological and thermal behaviors and chemical properties [7]. In addition, we measured transfer characteristic of a TFT structure prepared using and tried screen printing the resulting highly viscous IGZO precursor.

EXPERIMENT

Synthesis of precursor sols

Alcoholamines, Ethanolamine (EA), diethanolamine (DEA), and triethanolamine (TEA) as shown in Figure 1, were purchased from TCI. Indium nitrate trihydrate, gallium nitrate octahydrate, zinc nitrate hexahydrate, and urea were purchased from Wako Chemicals. All the chemicals were used without purification.

Typical synthesizing method for the IGZO precursors is as follow. Indium nitrate trihydrate (2.13 g, 0.006 mol), gallium nitrate octahydrate (0.40 g, 0.001 mol), and zinc nitrate hexahydrate (0.89 g, 0.003 mol) were dissolved in Milli-Q water (30 ml) and stirred in a beaker at room temperature in the air. Urea (6.01 g, 0.1 mol) was added and stirred in the metal nitrates aqueous solution. After dissolving urea completely, 30 ml of alcoholamine was added and white precipitation was appeared. Then, heating of the solution was started to 150 or 160 °C (in the case of TEA) during vigorous stirring and kept at 150 or 160 °C for 1 h during stirring. As the results, an opaque pale-orange paste (EA-precursor), a transparent pale-yellow sol (DEA-precursor), a the transparent dark-orange sol (TEA-precursor) were obtained from EA, DEA, and TEA, respectively. In the case of synthesis for the highly viscous DEA-precursor, indium nitrate trihydrate (10.7 g, 0.03 mol), gallium nitrate octahydrate (2.00 g, 0.005 mol), and zinc nitrate hexahydrate (4.45 g, 0.015 mol) were dissolved in Milli-Q water (50 ml) and stirred in a beaker. Urea (30.0 g, 0.5 mol) and 60 ml of DEA were added. The resulting product was a transparent orange sol.

Thin film preparation and screen printing

Thin IGZO film was prepared for measurement of transfer characteristic. The DEA-precursor (0.1 g) was diluted with ethylene glycol (0.3 g) and the diluted precursor was coated on a silicon dioxide (300 nm) surface of a silicon wafer with a spin coater at 3000 r.p.m. for 30 s. The resulting film was sintered in an electric oven at 450 °C for 30 min. Then, in order to prepare a thin film transistor structure, gold (200 nm) as a source and drain electrodes (channel width: 500 μm, channel length: 50 μm) was deposited on the surface of the sintered film through a metal mask by vacuum vapor deposition.

Screen printing was carried out on a silicon dioxide surface of a silicon wafer by hand using a stainless mesh (500 inch^{-1}) coated with 10 μm of a photosensitive emulsion layer with opening print patterns.

Measurement

Viscosity was measured with a rotational rheometer (Rheologia A300, ELQUEST). Thermogravimetry-differential thermal analysis (TG-DTA) was carried out with a thermal analyzer (Exstar TG/DTA 6200, SII). Transfer characteristic for the thin IGZO film was measured with a Semiconductor Characterization System 4200 (KEITHLEY) in nitrogen atmosphere.

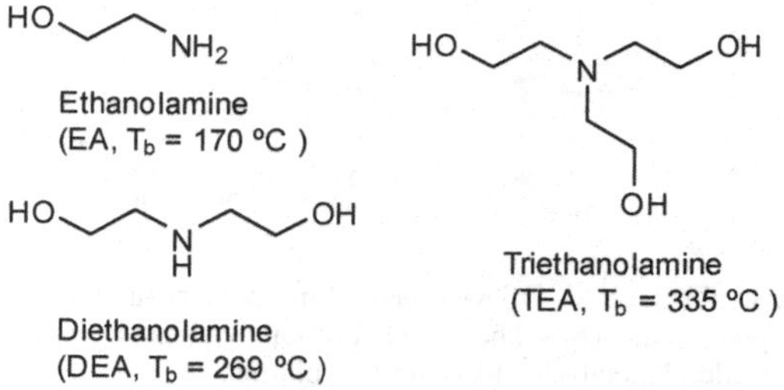

Ethanolamine
(EA, T_b = 170 °C)

Diethanolamine
(DEA, T_b = 269 °C)

Triethanolamine
(TEA, T_b = 335 °C)

Figure 1 Chemical structures and the boiling points (T_b) of alcoholamines.

DISCUSSION

Chemical state and viscosity of precursor sols

In the synthesis of the IGZO precursors, the metal hydroxides are obtained just after addition of alcoholamine in the existence of water at room temperature due to rise in the pH. Homogeneous hydroxylation of metal ions and formation of metal ammine complex [8] from metal hydroxides are given by addition of urea caused by thermal decomposition of urea to ammonia. In addition, alcoholamines chelate metal ions [7,9]. After addition of EA, the white precipitation is colored to pale-orange with a rise in temperature to 150 °C. In this case, metal hydroxides are kept in aggregational state with the size larger than the wavelength of visible light and partial metal ions would form the ammine complex, resulting in formation of slight-colored suspension. In the case of DEA and TEA, the white precipitation disappears gradually and the solutions become transparent and colored with increase in the temperature higher than 110 °C. In these cases, metal ions probably form ammine complex and/or chelating of alcoholamines homogeneously. Rheological properties reflect the chemical states of the IGZO precursors. In the case of the EA-precursor, increase in the shear rate brings about decrease in the viscosity as shown in Figure 2a. The pseudoplastic flow is the typical characteristic of suspension. The DEA- and TEA-precursors show Newtonian flow in the shear rate lower than 100 s^{-1} and pseudoplastic flow in the higher shear rate and these viscosities are higher than that of the EA-precursor. Pure DEA and TEA are originally viscous fluids. In addition, probably the DEA- and TEA-precursors form homogeneous networks attributed to hydrogen bonding between alcoholamine molecules coordinated with metal ions [7]. In the lower shear rate, the characteristic like pure DEA and TEA molecules, which are not associated with chelating metal ions, appears as Newtonian flow. In the higher shear rate, molecules in the sols would tend to be released from the networks. The highly viscous DEA-precursor includes higher concentration of the DEA molecules associated with chelating metal ions. Thus, the initial viscosity is higher and pseudoplastic flow appears even in the lower shear rate as shown in Figure 2b.

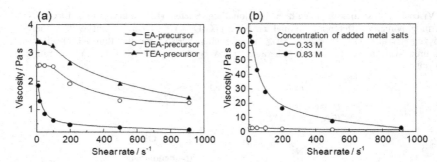

Figure 2 Viscosity curves for the EA-, DEA-, and TEA-precursors (a) and for the DEA- and the highly viscous DEA-precursors (b). The concentration values in (b) indicate the mol concentration (M) of the added metal salts to the DEA volume.

Thermal characteristics of precursor sols

TG and DTA curves for the EA-precursor show a broad endothermic peak with weight loss at 100-180 °C due to evaporation of residual water and EA as shown in Figure 3. Another endothermic peak with weight loss is also observed at 230 °C, which would be caused by dehydration of metal hydroxides and subsequent formation of metal oxide (IGZO). An exothermic peak with weight loss appears at around 280 °C attributed to oxidative decomposition of ammine complex and/or chelate structure and subsequent formation of IGZO. An exothermic peak at 460 °C is due to oxidative decomposition of organic residue. In the curves for the DEA-precursor, an endothermic peak with weight loss at 240 °C indicates evaporation of DEA. The influence of dehydration of metal oxides might be also overlapped in this peak. Subsequently a sharp exothermic peak at 300 °C shows oxidative decomposition of ammine complex and/or chelate structure resulting in formation of IGZO. A small exothermic peak at 450 °C is due to oxidative decomposition of organic residue. In the curves for TEA precursor, no endothermic peak with weight loss is observed in the measured temperature range because TEA is hard to evaporate. The first exothermic peak at 280 °C would be due to oxidative decomposition of ammine complex in the sol. The other exothermic peaks at the higher temperature probably show oxidative decomposition of organic residue.

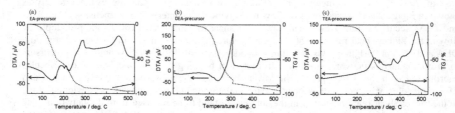

Figure 3 TG-DEA curves for the (a) EA-, (b) DEA-, and (c) TEA-precursors.

126

TFT characteristic of the IGZO film and screen printing

A thin film transistor structure was prepared using the DEA-precursor for measuring transfer characteristic of the IGZO thin film. According to the transfer characteristic as shown in Figure 4, the mobility and the on-off ratio of the IGZO thin film are estimated to be 0.02 $cm^2 \cdot V^{-1}s^{-1}$ and $\sim 10^5$. The transfer characteristic curve shows hysteresis. This suggests the existence of organic residue in the IGZO thin film even after sintering at 450 °C for 30 min. No specific peak in X-ray diffraction (XRD) measurement suggests the IGZO thin film is amorphous, though the detailed data is not shown here. Although the performance as TFT is still not good, the precursor has a potential as a semiconductor. This result triggers improvement of contents of the precursors, sintering methods and so on. Additionally, the highly viscous DEA-precursor shows a potential as an ink for screen printing with patterns as shown in Figure 5. The printed thickness is about 10 μm, depending on the thickness of the photosensitive emulsion layer with the opening print area coated on the mesh. In this case, the thickness is quite large as compared with that of spin-coated film. Currently, we are trying to microwave sintering method as one of the candidates for sintering of thicker patterns.

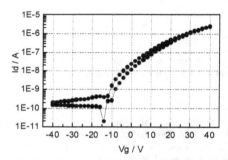

Figure 4 Transfer characteristic of the IGZO thin film prepared from the DEA-precursor.

100 mm

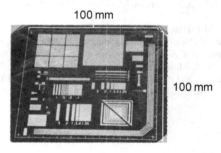

100 mm

Figure 5 Photograph of the patterns printed using the highly viscous DEA-precursor by screen printing on a silicon dioxide surface of a silicon wafer.

CONCLUSIONS

We synthesized viscous IGZO precursors using alcoholamines with primary-, secondary-, and tertiary amines by the simple method. EA is less ability to chelate metal ions as compared with DEA and TEA, because only the EA-precursor mainly includes aggregation of metal hydroxide, resulting in formation of the opaque paste. The DEA- and TEA-precursors seem to form homogeneous 3D-network of alcoholamines and metal ions with hydrogen bonding and chelating, leading to formation of the transparent sols. The higher concentration of adding metal salts derives from the higher viscosity and also pseudoplastic flow in the case of the DEA-precursor. Thermal analysis of the precursors suggests the suitable temperature for sintering as well as the chemical states. The sintered DEA-precursor indicates characteristic of a semiconductor in the TFT structure. The highly viscous DEA-precursor shows a potential as an ink for screen printing. The ideal precursor for screen printing has high viscosity and thixotropic nature. In addition, it desires to consist of decomposable materials at low temperature as possible. Screen printing must be a promising method for large-area printing and high-speed manufacturing of devices. Improvement of the contents of precursors and the sintering method is required for high-performance IGZO devices.

ACKNOWLEDGMENTS

We are grateful for technical supports by S. Manaka, Y. Kasuya, and N. Iwase in AIST. NF deeply thanks Dr. Y. Kusaka in AIST for valuable discussion on rheology.

REFERENCES

1. F. Garnier, R. Hajlaoui, A. Yassar, and P. Srivastava, *Science* **265**, 1684 (1994).
2. J. A. Rogers, Z. Bao, K. Baldwin, A. Dodabalapur, B. Crone, V. R. Raju, V. Kuck, H. Katz, K. Amundson, J. Ewing, and P. Drzaic, *Proc. Natl. Acad. Sci. USA* **98**, 4835 (2001).
3. Y. He, Z. Chen, Y. Zheng, C. Newman, J. R. Quinn, F. Dötz, M. Kastler, and A. Facchetti, *Nature* **457**, 679 (2009).
4. K. Nomura, H. Ohta, A. Takagi, T. Kamiya, M. Hirano, and H. Hosono, *Nature* **432**, 488 (2004).
5. S. K. Park, Y-H. Kim, and J-I. Han, *J. Phys. D: Appl. Phys.* **42**, 125102 (2009).
6. F. C. Krebs, M. Jørgensen, K. Norrman, O. Hagemann, J. Alstrup, T. D. Nielsen, J. Fyenbo, K. Larsen, and J. Kristensen, *Sol. Energy Mater. Solar Cells* **93**, 422 (2009).
7. R. Razali, A. K. Zak, W. H. A. Majid, and M. Darroudi, *Ceram. Int.* **37**, 3657 (2011).
8. T. Ramasami, R. K. Wharton, and A. G. Sykes, *Inorg. Chem.* **14**, 359 (1975).
9. P. C. Goh, K. Yao, and Z. Chen, *J. Phys. Chem. C* **116**, 15550 (2012).

Mater. Res. Soc. Symp. Proc. Vol. 1547 © 2013 Materials Research Society
DOI: 10.1557/opl.2013.1028

Structural and Functional Properties of Iron (II, III)-Doped ZnO Monodisperse Nanoparticles Synthesized by Polyol Method

Yesusa Collantes[1] and Oscar Perales-Perez[2]

[1] Department of Physics, University of Puerto Rico at Mayaguez, Mayaguez 00980, PR
[2] Department of Engineering Science and Materials, University of Puerto Rico at Mayaguez, Mayaguez, PR, 00680-9044

ABSTRACT

In this work, bare and (Fe^{3+} and Fe^{2+})-doped ZnO nanoparticles (NPs) have been synthesized in a polyol medium at 180°C. The synthesis in polyol allows a precise control of doping under size-controlled conditions. The Fe concentration varied in the 0-2 at. % range. As-synthesized samples were characterized by X-ray diffraction (XRD), Fourier Transform Infrared (FT-IR), Photoluminescence (PL) spectroscopy and Vibrational Sample Magnetometry (VSM). XRD measurements confirmed the formation of well crystallized wurtzite ZnO with absence of secondary phases in bare and doped samples; the average crystallite size was estimated at 8.4 ± 0.3 nm for bare ZnO NPs. Systematic shifts in the main diffraction peaks due to the incorporation of the dopant species were observed in the Fe^{3+} and Fe^{2+} doped-ZnO samples. FT-IR analyses evidenced the presence of organic moieties on the surface of the nanoparticles that are associated to the functional groups of polyol by-products; these adsorbed species could explain the observed stability of the NPs when suspended in water. PL measurements (excitation wavelength 345 nm) reveled that a tuning in the emission bands of ZnO NPs can be achieved through doping. VSM measurements evidenced a weak but noticeable ferromagnetic response at room temperature (RT) in doped samples.

INTRODUCTION

Zinc Oxide (ZnO) is an important wide band gap semiconductor (3.3 eV) with excellent optical properties. Hexagonal ZnO-wurtzite structure (a = 3.29 Å, and c = 5.24 Å), is transparent in the UV region and exhibits a large excitation binding energy at room temperature (60 meV) [1]. ZnO has been focus of many research in the last decades due to its applications in optics and optoelectronics [2,3]. Recently, a renewed interest of researchers in ZnO at the nanoscale is based on the fact that Zinc oxide (ZnO) nanoparticles (NPs) display unique chemical and physical properties and numerous potential applications. Doping of ZnO NPs also has been explored as an attempt to enhance and/or tune their functional properties and enable these NPs to find novel applications not only in spintronics [4], but also in biology and medicine, including biological fluorescence labeling, imaging, diagnosis and photo-dynamic therapy as theranostics materials [5][6]. ZnO is a likely material for biological applications since it is non-toxic, biodegradable [7,8], presents high thermal and chemical stabilities [9]. Although the doping of ZnO with different transition metal ions has been reported elsewhere, the effect of the dopant oxidation state on the materials properties at the nanoscale has not been altogether addressed yet. Accordingly, the present research is focused on the controlled synthesis and characterization of bare and Fe^{2+} and Fe^{3+} doped ZnO NPs through a modified polyol-based approach. The polyol acts as the solvent for the formation of NPs while enhancing their stability in water [10,11]. The corresponding structural, optical and magnetic properties were discussed as a function of the dopant concentration in the 0-2 at.% range.

EXPERIMENTAL

A. Materials

Iron (II) acetylacetonate [Fe(acac)$_2$, 99.5%], Iron (III) acetylacetonate [Fe(acac)$_3$, 99%] and zinc acetate dehydrate [Zn(OOCCH$_3$)$_2$·2H$_2$O, 99%] were precursors salts. Triethylene glycol (TREG, 99%) was solvent. The concentration was kept constant at 0.08mol/L in all experiments.

B. Synthesis of Bare and Fe(II and III)-doped ZnO Nanoparticles

The synthesis of bare ZnO NPs took place by heating a solution of Zinc acetate in TREG until 180°C under reflux and vigorous stirring for 30 minutes. The suspension was then cooled down to room temperature, yielding a cream suspension containing the ZnO NPs. In the synthesis of Fe^{3+}-doped ZnO, appropriate amounts of Fe(acac)$_3$ and Zinc acetate were mixed in TREG according to Zn$_{(1-x)}$Fe^{3+}$_{(x)}$O stoichiometry and heated up until 180°C under similar conditions as in bare ZnO. The colloidal suspensions of pure and doped ZnO nanoparticles obtained were centrifuged at 8,000 rpm, washed with ethanol three times and re-dispersed in water. The NPs recovered by centrifugation were dried at 60 °C for 24 hours and characterized. A similar procedure was carried out in the case of Fe^{2+}-doped ZnO NPs but using Fe(acac)$_2$ as the iron source. The dopant concentration was in the range of 0 to 2 at. %.

C. Characterization

Powder x-ray diffractograms were recorded using a Siemens D500 diffractometer (Cu-Kα radiation). Average crystallites sizes were calculated with Scherrer's equation [12]. Materials surface was studied through FT-IR spectra using a Shimadzu IR-Afinity spectrometer. As-synthesized powders were examined with a JEOL JEM-2200FS high resolution transmission electron microscope equipped with an x-ray energy dispersive spectrometer. PL measurements were carry out with a Fluoromax2 Photo-spectrometer (λ_{exc}=345 nm). RT hysteresis loops were recorded using a Lake Shore 7410 vibrational sample magnetometer.

RESULTS AND DISCUSSION

A. X-ray diffraction (XRD)

Figure 1-a displays XRD patterns of bare and Fe^{2+}-doped ZnO NPs (0, 0.5,1 and 2 at.%). All the diffraction peaks correspond to hexagonal wurtzite ZnO without secondary phases. The absence of impurity phases suggests the actual incorporation of Fe^{2+} (0.75 Å) in the Zn^{2+} (0.72 Å) sites. The average crystallite size for bare ZnO NPs was estimated at 8.4 ± 0.3 nm; it was very similar for the Fe^{2+}-doped samples (8.5 ± 03 nm, 6.2 ± 0.7 nm, 7.0 ± 0.2 nm for the NPs doped at 0.5, 1 and 2 at.%, respectively). Figure 1-b shows the detail of the XRD patterns of bare and Fe^{2+}-doped NPs; the systematic shift to the left of the peaks (002) and (101) with increasing dopant at.% evidences the actual incorporation of the Fe^{2+} ions into the ZnO lattice host. This shift could be explained in terms of the distortion (expansion) of the lattice due to the substitution of ions with different ionic radii. Figure 1-c presents XRD patterns of bare and Fe^{3+}-doped NPs (0, 0.5, 1 and 2 at.%). All the diffraction peaks correspond to well crystallized wurtzite ZnO without impurity phases, which suggest the actual substitution of Fe^{3+} (0.69 Å) in the Zn^{2+} (0.72 Å) sites. The corresponding average crystallite sizes were 5.9 ± 06 nm, 6.0 ± 0.3 nm and 6.3 ± 0.2 nm for 0.5, 1 and 2 at. %, respectively; Fe^{3+} ions degrade the wurtzite structure retarding the crystallization process. The same behavior has been observed in the doping of ZnO with other trivalent ions [13,14]. Figure 1-d displays the detail of the XRD patterns of bare and Fe^{3+}-ZnO NPs; in this case, a noticeable and systematic shift to larger diffraction angles with the increase of the dopant concentration was observed. This trend reinforces the hypothesis of an actual substitution of Zn^{2+} by Fe^{3+}.

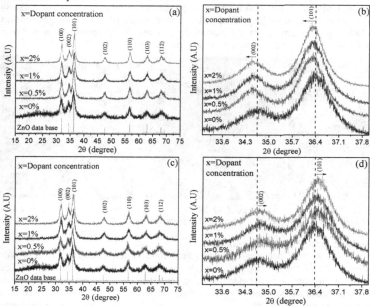

Figure 1. XRD patterns of Fe^{2+}-doped ZnO NPs synthesized at various dopant atomic percentages, 'x', (a); detail of the XRD patterns of Fe^{2+}-ZnO NPs, (b); XRD patterns of Fe^{3+}-doped ZnO NPs synthesized at dopant concentration 'x' in the 0-2 at.% range, (c); detail of the XRD patterns of Fe^{3+}-ZnO NPs, (d).

B. Fourier Transform Infrared (FTIR) spectroscopy

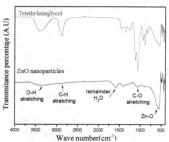

Figure 2. FTIR spectra of pure TREG and bare ZnO NPs.

Figure 2 shows the FTIR spectra of bare ZnO NPs and pure TREG. The spectrum of ZnO NPs reveals the presence of adsorbed byproducts that are related to the polyol functional groups, which is evidenced by the presence of the bands at 2,925-2,809 cm^{-1} (C–H stretching, ester group) and 1,116 – 1,050 cm^{-1} (C–O stretching, alcohol group) [10]. The broad band centered at 3,351 cm^{-1} represents O–H stretching vibrations related to surface hydroxyl groups [10]. The Zn-O bond was evidenced by the intense band at 585 cm^{-1} [15]. The band at 1,630 cm^{-1} indicates the presence of residual water (H$_2$O) in the sample [16]. Therefore, the water stability observed can be attributed to the 'steric' repulsion due to a positive charged surface generated by the adsorption of polyol-related species onto the NPs surface [17].

C. High Resolution Transmission electron microscopy (HRTEM) analysis

The samples were analyzed by HRTEM technique. Figure 3 is a montage of representative images: bare ZnO NPs of ~8 nm with a FTT as inset, (a); an overview of monodisperse single crystals of ZnO, (b); and another ZnO NP of ~7.6 nm in diameter, (c). These images confirm the high crystallinity of the samples at the nanoscale and are in agreement with the average crystallite size calculated from the x-ray diffraction analyses.

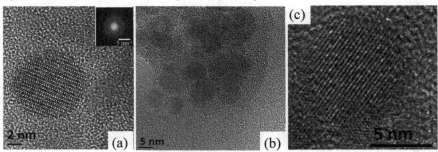

Figure 3. HRTEM images of: (a) a bare ZnO NP of ~ 8nm, the inset is a FFT; (b) overview of monodisperse ZnO NPs; (c) ZnO NP of ~ 7.6 nm with a plane spacing of 3 Å.

D. Photoluminescence (PL) Measurements

Figure 4-a presents the PL spectra of bare and Fe^{2+}-doped ZnO NPs (0, 0.5 y 2 at.%). The spectrum of bare ZnO exhibits the well-known UV emission of ZnO at 362 nm (3.42 eV) that is attributed to the recombination of the electron-holes pairs, and the green emission of ZnO at 554 nm related to superficial defects and oxygen or Zinc vacancies [1][18][19]. The presence of other visible emissions, like the one at 390 nm, is very interesting. Violet emissions have been detected in small ZnO NPs and were assigned to existence of zinc interstitials (Zn_i) [20]. The very intense 390 nm-peak observed in the as-synthesized NPs can enable a novel and simple polyol-synthesis route to produce ZnO NPs exhibiting strong violet luminescence. In turn, the small band at 380 nm can be attributed to transitions from Hydrogen interstitials (H_i) levels in ZnO, which usually overlap the Zn_i levels [21]. For the Fe^{2+}-doped samples, it was observed a quenching-by-concentration effect in the UV emission. This quenching occurs when the dopant concentration is so high that the probability of non-radiative transitions exceeds that of emission[22]. Besides, the ratio of intensities of the two main emission peaks, $I_{362\ nm}/I_{390\ nm}$, was higher for bare NPs than for the 0.5 at. % doped sample. In the later, the predominance of the 390 nm peak (violet emission) intensity became evident. Both, 362 nm and 390 nm emissions were quenched at larger dopant concentration, suggesting the doping threshold[23] for Fe^{2+}-doped ZnO samples. The predominance of the violet emission could be related to the increase of donor carriers that increase the probability to fill those Zn_i sates. Figure 4-b shows the spectra of bare and Fe^{3+}-doped ZnO NPs (0, 0.5 and 2%). As observed, the Fe^{3+} ions produced a weaker quenching in the 362nm-UV emission intensity compared to the Fe^{2+}-doped system. Actually, the intensity of this peak was not quenched in the 0.5 at. % doped sample; a similar behavior was reported in other systems including trivalent dopants [13,14]. The $I_{362\ nm}/I_{390\ nm}$ ratio did not change as drastically as in the Fe^{2+} doped-sample. The green emission vanished in all the doped samples suggesting that some of the dopant species are filling the Zn vacancies related with this emission. [13,14]

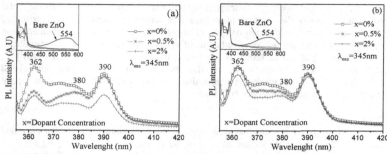

Figure 4. (a) PL spectra of bare and Fe^{2+}-ZnO NPs synthesized at various dopant atomic percentages, 'x'; (b) PL spectra of bare and Fe^{3+}-ZnO NPs synthesized at 'x' values. The insets show the complete spectra.

E. Magnetic Measurements at Room-Temperature

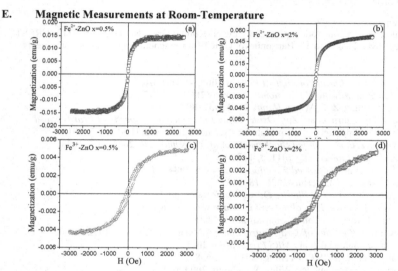

Figure 5. M-H curves of Fe^{2+}-doped ZnO NPs: x=0.5 at. %, (a); x=2 at.%, (b). Magnetization curves of Fe^{3+}-doped ZnO NPs: x=0.5 at.%, (c); x=2 at.%, (d).

The M-H curves of Fe^{2+}-doped ZnO NPs with x=0.5 and 2 at. % are shown in Figure 5 (a-b). The saturation magnetization varied from 15 $m$emu/g to 52 $m$emu/g. The increase in magnetization is attributed to the major concentration of available spins in the sample with the higher dopant concentration. The samples did not exhibit noticeable coercivity. Figure 5 (c-d) displays the magnetization curves of Fe^{3+}-doped ZnO NPs (x=0.5 at.%, and 2 at.%, (d). The saturation magnetization was 5.3 $m$emu/g and the coercivity 104 Oe for the Fe^{3+}-doped ZnO at x= 0.5 at. %. In this case, the Fe^{3+}-doped ZnO (x= 2 at.%) did not show saturation although a coercivity of

73 Oe was measured. As evidenced, all samples exhibited a weak but noticeable ferromagnetic behavior at RT while retaining their optical properties. The ferromagnetism might be attributed to the super exchange interaction through oxygen (TM-O_2-TM) or to the exchange interaction between spins of the band carriers and localized spins of the Fe ions. Moreover, a defect-induced ferromagnetism can be explained in terms of the bound magnetic polarons model (BMP)[24], which is in agreement with our PL measurements that showed a strong violet emission related with structural defects (Zn_i). This weak ferromagnetism reinforces the hypothesis of an actual incorporation of the dopant species into the ZnO host.

CONCLUSION
A novel polyol-based synthesis process was developed to produce violet luminescent pure and doped ZnO NPs. The analysis of the oxidation state of the dopant suggest that doping of ZnO with Fe^{3+} ions display better optical and magnetic responses. These NPs can find potential applications in biology and medicine such as biological fluorescence labeling, imaging, diagnosis and cancer therapy.

Acknowledgments
This material is based upon work supported by the NSF under Grant No. HRD 0833112 (CREST program at UPRM). Thanks to Ms Barrionuevo, UPRRP for magnetic properties measurements.

References
[1]   U. Özgür, et al., *Journal of Applied Physics* **2005**, *98*, 041301.
[2]   H.-K. Fu, et al, *Advanced Functional Materials* **2009**, *19*, 3471.
[3]   Y. Qin, X. Wang, Z. L. Wang, *Nature* **2008**, *451*, 809.
[4]   C. Franz Klingshirn, et al., *Zinc Oxide: From Fundamental Properties Towards Novel Applications*; Springer Series in Materials Science 120: Berlin Heidelberg, **2010**.
[5]   A. Fernandez, et al., *Applied Biochemistry And Biotechnolog0y* **2011**, *165*, 1628.
[6]   T. Nann, *Nano Biomed Eng* **2011**, *3*, 137.
[7]   Z. L. Wang, et al., *Advanced Functional Materials* **2004**, *14*, 943.
[8]   J. Zhou, *Advanced Materials* **2006**, *18*, 2432.
[9]   E. Çetinörgü, *Journal of Physics D: Applied Physics* **2007**, *40*, 5220.
[10]   W. Cai, et al., *Journal of colloid and interface science* **2007**, *305*, 366.
[11]   E. V. Panfilova,et al., *Colloid Journal* **2012**, *74*, 99.
[12]   B. D. Cullity, *Elements of X-ray Diffractions*; Addison Wesley, MA, **1972**; p. 102.
[13]   M. a. Gálvez Saldaña, et al., *MRS Proceedings* **2011**, *1368*, 2.
[14]   M. Galvez, et al., *MRS Proceedings* **2010**, *1256*, 8.
[15]   R. Cornell, *The iron oxides*; Wiley, Weinheim, **2003**; p. 141.
[16]   T. Rajh, et al., *Physical Chemistry B* **2002**, *106*, 1053.
[17]   J. Zhao, et al., *American Ceramic Society* **2011**, *93*, 725.
[18]   L. Feng, et al., *Applied Physics Letters* **2009**, *95*, 053113.
[19]   E.-S. Jeong, et al., *Journal of Nanoscience and Nanotechnology* **2010**, *10*, 3562.
[20]   V. Ischenko, et al., *Advanced Functional Materials* **2005**, *15*, 1945.
[21]   D. Hofmann, et al., *Physical Review Letters* **2002**, *88*, 045504.
[22]   S. Shionoya, W. M. Yen, *Phosphor Handbook*; CRC Press, Ed.; Phosphor Research Society: Boca Raton, Florida, **1997**.
[23]   H. Morkoç, U. Özgür, *Zinc Oxide;* Wiley-VCH, **2008**.
[24]   J. M. D. Coey, et al., *Nature materials* **2005**, *4*, 173.

Mater. Res. Soc. Symp. Proc. Vol. 1547 © 2013 Materials Research Society
DOI: 10.1557/opl.2013.638

Size-Controlled Synthesis of MgO Nanoparticles and the Assessment of Their Bactericidal Capacity

Yarilyn Cedeño-Mattei[1,2], Myrna Reyes[1], Oscar Perales-Pérez[1,2], and Félix R. Román[1]

[1]Department of Chemistry, University of Puerto Rico, Mayaguez, PR 00681-9000, U.S.A.
[2]Department of Engineering Science and Materials, University of Puerto Rico, Mayaguez, PR 00681-9000, U.S.A.

ABSTRACT

The present work focuses on the development of a reproducible and cost-effective size-controlled synthesis route for nanoscale MgO and the preliminary assessment of its bactericide capacity as a function of crystal size. Nanoscale MgO was produced through the thermal decomposition of Mg-carbonate hydrate precursor (hydromagnesite) synthesized in aqueous phase. The exclusive formation of the MgO phase, with an average crystallite size between 7 and 13 ± 1 nm, was evidenced by X-Ray Diffraction and HRTEM analyses. Fourier Transform – Infrared spectroscopy confirmed the evolution of the precursor into the desired MgO structure. The bactericidal tests were conducted by measuring the optical density at 600 nm of E. coli in presence of MgO nanoparticles of specific sizes. MgO nanocrystals with average crystallite sizes of 13nm inhibited bacterial growth up to 35% at 500 mg MgO/L. The mechanism of inhibition could be attributed to the formation of superoxide species on the MgO surface.

INTRODUCTION

MgO has been used in catalysis, thermal insulating, and adsorption, among other applications [1-2]. The antimicrobial activity of environmental-friendly and chemically stable MgO has also been suggested for food packaging applications. A major concern for the food industry is the development of more efficient and effective materials for food preservation and protection against human health-compromising microorganisms. Earlier works have been focused on the evaluation of carbon nanotubes, silver, copper, zinc, ZnO, CuO, NiO, Sb_2O_3, TiO_2, gold, colloidal clays and zeolite minerals as anti-fungal or anti-microbial materials [3-6]. Compared to others, MgO is expected to provide an improved food packaging solution under cost-effective conditions. Also, the capability of nanoscale MgO to destructively adsorb chemical warfare agents and its bactericidal activity even in absence of UV illumination, coupled with its ability to block UV light, enable this nanomaterial to be considered a promising candidate for food protection and preserving applications. Despite of the remarkable features of MgO, there is still a lack of systematic research on its size-controlled synthesis, which could modify its corresponding size-dependent properties at the nanoscale. On this basis, the present work addresses the size-controlled synthesis of MgO nanocrystals, their structural and morphological characterization, and the preliminary assessment of the bactericidal capacity in presence of E. coli.

EXPERIMENTAL

Materials

All regents were of analytical grade and used without any further purification. $Mg(NO_3)_2$ (ACS, 98-102%, Alfa Aesar), Na_2CO_3 (≥ 99%, Sigma-Aldrich), and NaOH (pellets, 98%, Alfa Aesar) were used for the synthesis of the magnesium carbonate hydroxide precursor.

Synthesis of Magnesium Hydroxide Carbonate Precursor and MgO Nanocrystals

The synthesis method employed in the formation of magnesium carbonate hydroxide (hydromagnesite) precursor is a modification of the one proposed by Y. Zhao *et al* [7], involving separate nucleation and aging steps in aqueous phase. Solution-1 consisted of 100 mL 0.3 M $Mg(NO_3)_2$, while solution-2 contained 100 mL of stoichiometric amounts of Na_2CO_3 and NaOH. Solutions 1 and 2 are added simultaneously into the reaction vessel and homogenized at 11000 rpm for 2 minutes (nucleation step). The resulting slurry is stirred and heated at 100°C for different reaction times (aging step). At the end of the aging time, the precipitate was washed three times with deionized water and dried at 50 °C for 24 hours. The formation of the hydromagnesite is explained by the following reaction.

$$5Mg(NO_3)_2 + 4Na_2CO_3 + 2NaOH \xrightarrow{\Delta} 4MgCO_3 \cdot Mg(OH)_2 \cdot 4H_2O \qquad (1)$$

The synthesized magnesium carbonate hydroxide precursor was thermally treated in order to promote the MgO formation. This is a two-step process based in the dehydroxilation and decarbonation of the precursor, which leads to:

$$4MgCO_3 \cdot Mg(OH)_2 \cdot 4H_2O \xrightarrow{\Delta} 5\ MgO + 5\ H_2O + 4\ CO_2 \qquad (2)$$

Nanocrystals Characterization

The crystalline structure of hydromagnesite and MgO were investigated using a Siemens D500 X-Ray Diffractometer with Cu K_α radiation. The average crystallite size was calculated using the Scherrer's equation [8]. A Shimadzu IRAffinity-1 Fourier Transformed Infrared Spectrophotometer was used to confirm the formation of the desired structures. The morphology and size of the materials were examined using a JEM-ARM200cF Transmission Electron Microscope.

Assessment of the MgO Bactericidal Capacity

The capacity for inhibition of bacterial growth was assessed in presence of MgO nanocrystals with average crystallite sizes around 7 and 13nm. MgO concentrations of 250, 350, and 500 mg/L were evaluated. A fixed volume of 50 µL of the inoculum of *E.coli* at a concentration of 10^8 cells/mL was mixed with the different concentrations of nanocrystals and the final volume was filled up to 5 mL by using Tryptic Soy Broth. A negative control was examined in absence of MgO. The bacterial growth was assessed by measuring the optical density at 600 nm at one hour-intervals using a Spectronic 20 D+ Spectrophotometer.

DISCUSSION

XRD Analyses

Figure 1-a shows the XRD patterns of the powders of magnesium carbonate hydroxide precursor synthesized at different reaction times. Broad and noisy diffraction peaks were observed in the pattern corresponding to the sample synthesized at 0 hours of reaction. This pattern corresponds to the sample that was recovered right after the nucleation stage, i.e. without any aging. The prolonging of the aging time from 0.5 hours up to 4 hours was conducive to the formation of well crystallized magnesium carbonate hydroxide (hydromagnesite) precursor. Reaction times longer than 4 hours resulted in the predominant formation of magnesium hydroxide (brucite). On this basis, the one-hour aged hydromagnesite sample was thermally treated in air at 600 °C; the corresponding XRD patterns are shown in Figure 1-b. The evolution from hydromagnesite to the desired oxide structure was monitored in the 15-60 minutes of heating range. The presence of diffraction peaks corresponding to the crystallographic planes (111), (200), and (220) of cubic MgO-periclase became evident. The lattice parameter for the sample synthesized after one hour of thermal decomposition was estimated at 4.22 Å, which is in good agreement with the bulk value for MgO (4.21 Å) [9]. In general, the average crystallite sizes of MgO synthesized from the hydromagnesite aged for different times ranged between 7 and 13 ± 1 nm. MgO nanocrystals produced from highly crystalline precursors (e.g. brucite produced after 8 hours of aging) led to smaller crystal sizes than those produced from poorly crystalline ones (e.g. non-aged hydromagnesite). This trend could be attributed to the fact that highly ordered structures, like brucite, would demand more energy to disturb the original structure and cause the atomic rearrangement associated to the formation of the MgO phase, and subsequent crystal growth. On the contrary, less crystalline structures (non-aged hydromagnesite) would require less thermal energy to be converted into the oxide phase and could take advantage of the excess of energy to promote crystal growth.

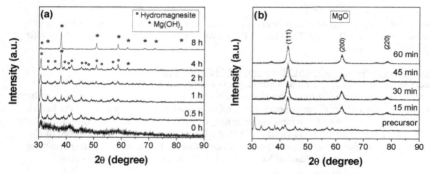

Figure 1. XRD patterns of: (a) hydromagnesite and/or brucite precursors synthesized at different reaction (i.e. aging) times; (b) MgO powders synthesized at 600 °C and different heating times. The precursor corresponds to the hydromagnesite aged for one hour.

Surfactants such as sodium oleate (Na-oleate), sodium dodecyl sulfate (SDS), and cetyltrimethylammonium bromide (CTAB), in addition to the polymer polyvinylpyrrolidone (PVP), were also evaluated during the synthesis of the precursors in order to avoid or minimize particles agglomeration. The XRD patterns corresponding the precursors produced in presence of the mentioned surfactants and/or PVP at a [surfactant or polymer]/[Mg^{2+}] molar ratio of 0.1, are shown in Figure 2-a. As seen, there is no remarkable difference in the crystallinity of the as-synthesized powders. These precursors were thermally treated in air at 600 °C. The average crystallite sizes of produced MgO ranged between 9-11 nm, 8-10 nm, 7-9 nm, and 8-10 nm for the samples produced in presence of Na-oleate, SDS, CTAB, and PVP, respectively. Comparatively speaking, the presence of Na-oleate led to the formation of bigger crystallites. Accordingly, the synthesis of the precursor phase was evaluated at different [Na-oleate]/[Mg^{2+}] molar ratios in the 0.1-1 range; the corresponding XRD patterns are shown in Figure 2-b. An increase in the [Na-oleate]/[Mg^{2+}] ratio was conducive to the formation of poorly crystalline hydromagnesite. As expected from the considerations regarding the effect of the precursor crystallinity on the oxide crystallite size, the average crystallite sizes ranged between 10-11 nm, 11-13, and 10-13 nm when the MgO was produced by heating the precursors synthesized at [Na-oleate]/[Mg^{2+}] molar ratios of 0.1, 0.5, and 1, respectively.

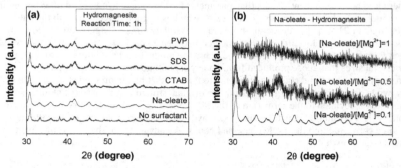

Figure 2. XRD patterns corresponding to: (a) hydromagnesite precursors synthesized in presence of surfactants (Na-oleate, SDS, and CTAB) and PVP; (b) hydromagnesite precursors synthesized at different [Na-oleate]/[Mg^{2+}] ratios.

Fourier Transformed - Infrared Spectroscopy Measurements

Figure 3 shows the FT-IR spectra of hydromagnesite produced at a [Na-oleate]/[Mg^{2+}] molar ratio of 0.5 and the MgO resulting after thermal treatment for 1 hour at 600 °C in air. The sharp bands at 1475 and 1420 cm^{-1} corresponds to the CO$_3^{2-}$ asymmetric stretching while the band at 1100 cm^{-1} is attributed to CO$_3^{2-}$ symmetric stretching of hydromagnesite. The other strong band at 3658 cm^{-1} is attributed to O – H vibration [10]; the C – H stretching band, associated to Na-oleate groups, was also detected in the 2962-2853 cm^{-1} range. The spectrum corresponding to MgO clearly shows an intense band at 540 cm^{-1} that corresponds to the Mg-O vibration.

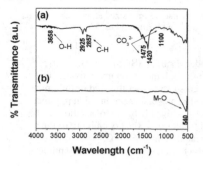

Figure 3. FT-IR spectra of hydromagnesite precursor synthesized in presence of Na-oleate, (a), and the MgO resulting after thermal treatment of the same precursor at 600 °C, (b).

Transmission Electron Microscopy Analyses

TEM and HRTEM images of 13 nm MgO are shown in Figure 4. The nanometric nature of the MgO particles and their high crystallinity was evidenced (Figures 4 a-b). Figure 4-c corresponds to the MgO produced from starting $Mg(OH)_2$, brucite, precursor. The platelet-shaped particles actually consists of aggregates of extremely small nanocrystals that are clearly observed in the HRTEM image of figure 4-d.

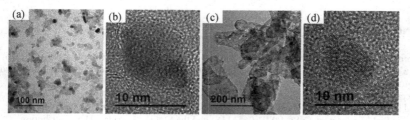

Figure 4. (a) TEM image of 13nm-MgO nanocrystals; (b) HRTEM image showing the crystallographic planes of a 13 nm-MgO crystal; (c) TEM image of MgO platelets produced from brucite precursor; (d) detail of a single 7 nm-MgO crystal included within the platelets of image (c).

Assessment of the Bactericidal Capacity of MgO

The bacterial growth curve of *E. coli* in presence of different concentrations of MgO nanocrystals and 4 hours of contact are presented in Figure 5. Concentrations of 250, 350, and 500 mg/L of MgO were considered. The corresponding percentages of growth inhibition varied between 17% - 35% and 14% - 24% when 13 nm- and 7 nm MgO nanoparticles were used, respectively. The bactericidal capacity of MgO can be attributed to the generation of superoxide species on the MgO surface [11]. On this basis, it could be expected that smaller MgO should represent larger specific surface area and hence a higher intrinsic bacterial growth inhibition capacity; however, the observed bacterial growth inhibition was higher for the larger (13nm) crystals. This trend could be attributed to the platelet-shape of the MgO aggregates that would exhibit less specific surface area available for the generation of the superoxide species.

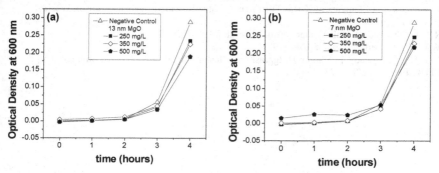

Figure 5. Bacterial growth curves for *E. coli* in presence of: (a) 13 nm, and (b) 7 nm MgO at different concentrations. The negative control tests correspond to the bacterial growth in absence of MgO.

CONCLUSIONS

MgO nanocrystals were successfully synthesized in the 7 – 13 nm range. The formation of the desired structure was confirmed by X-Ray Diffraction and FT-IR spectroscopy techniques. The corresponding bactericidal capacity was dependent of the crystallite size and particles concentration. An inhibition of bacterial growth up to 35% was obtained for 13 nm MgO at 500 mg/L.

ACKNOWLEDGMENTS

This project is supported by the Agriculture and Food Research Initiative Competitive Grant No. 2012-67012-19806 from the USDA-National Institute of Food and Agriculture. The authors also thank to the USDA-NIFA Center for Education and Training in Agriculture and Related Sciences (CETARS), Competitive Grant No. 2011-38422-30835. The TEM facility at FSU is funded and supported by the Florida State University Research Foundation, National High Magnetic Field Laboratory (NSF-DMR-0654118) and the State of Florida. The contribution from Professor Alicia Moradillos, Director of the Science and Technology Department, Antillean Adventist University at Mayaguez is also acknowledged.

REFERENCES

1. T. Selvamani, A. Sinhamahapatra, D. Bhattacharjya, I. Mukhopadhyay, Mat. Chem. Phys. **129**, 853 (2011).
2. Z. Gao, L. Wei, T. Yan, M. Zhou, Appl. Surf. Sci. **257**, 3412 (2011).
3. R. Kumar and H. Münstedt, Biomaterials **26**, 2081 (2005).
4. S.Y. Liau, D.C. Read, W.J. Pugh, J.R. Furr, A.D. Russell, Lett. Appl. Microbiol. **25**, 279 (1997).
5. Q. Li, S. Mahendra, D.Y. Lyon, L. Brunet, M.V. Liga, D. Li, P.J.J. Alvarez, Water Res. **42**, 4591 (2008).
6. Y.-W, Bae and Y.-J. Ann, Sci. Total Environ. **409**, 1603 (2011).
7. Y. Zhao, F. Li, R. Zhang, D.G. Evans, X. Duan, Chem. Mater. **14**, 4286 (2002).
8. B. D Cullity, in Elements of X-Ray Diffraction, edited by Morris Cohen (Addison Wesley, MA, 1972), p. 102.
9. R.T. Downs and M. Hall-Wallace, Am. Mineral. **88**, 247 (2003).
10. W.B. White, Am. Mineral. **56**, 46 (1971).
11. J. Sawai, E. Kawada, F. Kanou, H. Igarashi, A. Hashimoto, T. Kokugan, M. Shimizu, J. Chem. Eng. Jpn. **29**, 627 (1996).

Mater. Res. Soc. Symp. Proc. Vol. 1547 © 2013 Materials Research Society
DOI: 10.1557/opl.2013.854

Synthesis of Urethane Base Composite Materials with Metallic Nanoparticles

Anayansi Estrada Monje [1], J. Roberto Herrera Reséndiz [2]
[1] CIATEC, A.C. 201 Omega, Industrial Delta, León, Guanajuato, México, 37545.
[2] A. P Resinas, Calle Norte 4 No.3, Nuevo Parque Industrial San Juan del Río, Querétaro, México, 76809

ABSTRACT

The antimicrobial properties of polymer materials are used in a verity of applications. Silver nanoparticles are commonly applied to polyurethane foams to obtain antifungal properties. For this study a series of nanocomposites (PU–Ag) from a urethane-type polymer (PU) were reinforced with various amounts of silver nanoparticles having an average size of 20 nm. The surface morphology and antifungal capacity of the nanocomposites were evaluated. As a result, a different surface morphology from PU was found in PU–Ag nanocomposites. The latter nanocomposite showed enhanced thermal and mechanical properties, when compared with the PU without silver nanoaprticles. The nanocomposite also exhibited good antifungal properties that can be used in a variety of applications.

INTRODUCTION

Polyurethane (PU) polymer is commonly used in a variety biomedical application due to its biocompatibility, easy processing and positive physical-mechanical properties [1-3]. PU is comprised of rigid and soft segments enabling its final properties to be manipulated by altering the ratio or chemistry of the batch components. [4]. The term nanonomposite is used to describe polymeric systems that contain nanometric reinforcements, with an average particle size less than 100 nm, dispersed into a polymer matrix [5, 6]. These nanomaterials have gained interest in the recent years due to their unique properties [7]. Presently, the synthesis of reinforced polymer with metallic nanoparticles has fostered innovative ways of engineering materials that exhibit better electrical, optical, mechanical, antifungal and antibacterial properties.

The antifungal properties have been familiar to scientists for decades [8], with silver derivatives like sulfadiazines used to prevent bacteria growth in potable water bottles [9, 10]. For example, silver derivatives have been applied in a polypropylene matrix for sanitary applications such as surgical masks, filters, diapers, etc. [11]. Although the antibacterial effect of silver on microorganisms is well known, the mechanism by which this effect is achieved is only partially understood. Scientists have hypothesized that silver strongly interacts with thiol groups of particular enzymes, by inactivating the enzyme and inhibiting vital processes within the microorganisms [11, 12]. Other studies have demonstrated structural changes in the cellular membrane, as well as, the formation of small electron-dense granules when combined with silver and sulfur. An alternative mechanism involves the absorption and accumulation of silver ions (Ag^+) in the bacterial cells causing shrinkage of the cytoplasm or the detachment of the cellular wall; as a result, the DNA condenses and loses it replication ability.

In addition, silver nanoparticles have also demonstrated an inhibiting effect on viruses. [15]. There are fewer reports regarding the effects of silver nanoparticles on fungi. However, one existing example of example which demonstrates the antifungal properties of silver nanoparticles is found in the work of Keul-Jun Kim [16]; there the research was shown the antifungal activity of silver nanoparticles against *Trichophyton mentagrophytes* (*T. mentagrophytes*) and some Candida species. In this experiment, a polyurethane composite material with different silver nanoparticles concentrations was combined during the raction injection molding process (RIM). Then composite material's antifungal activity was probed against *T. mentagrophytes*, and the effect of the silver nanoparticles on the composite's physical-mechanical and thermal properties were studied. In the present work, the presence of silver nanoparticles in the polyurethane composites has also demonstrated an inhibiting effect against *T. mentagrophytes* fungus.

EXPERIMENTAL

Polyurethane synthesis

PU in this study was synthesized during the reaction injection molding process. For this experiment a system consisting in a polyester-like polyol with a tin catalyst (stream A) and isocyanate (stream B) were used.

Nanocomposites of polyurethane and silver

The synthesis of nanocomposite materials was batched following a similar procedure as the one used for PU synthesis with one exception. The exception in this process was that in stream "A" included different concentrations of nanosilver dispersions. The concentrations of nanosilver into polyurethane used to synthesize the composite materials were 0.000569, 0.001420, 0.002840 and 0.005690 by weight percent.

Morphology, thermal properties and storage modulus of the nanocomposites

A thermogravimetric analyzer, TA Instruments TGA 2050, was used to perform the thermogravimetric analysis (TGA) with a heating speed of 10 °C/min under nitrogen with 5 mg of sample material. Morphology, microstructure and the nanoparticles dispersions in the polymer matrix of the composite materials were studied using a Scanning Electronic Microscope. Dynamic mechanic analysis was made in a TA Instruments 2980 analyzer with 1 Hz of frequency, with a heating speed of 5 °C/min and a temperature range of -100 to 100 °C.

Inhibition of Fungi growth

Specific growth media for *T. Mentagrophytes* fungus was used in the petri dishes. 1 cm^2 samples of PU and the series of PU nanocomposites were placed into inoculated petri dishes with *T. Mentagrophytes*. The inoculated petri dishes with the samples were then incubated at 30 °C for 168 hours. The presence or lack of an inhibition halo in the petri dishes was noted at the conclusion of the experiment.

DISCUSSION

Nanocomposites characterization

Table 1 identifies the composite materials and the different concentrations of silver nanoparticles in, Table 1 also includes the storage modulus of the composite materials at different temperatures, as well as, the decomposition temperatures Td1 for the soft segments and Td2 for rigid segments of PU. Figure 1 shows the thermogravimetric measurement of the PU without silver nanoparticles and the PU nanocomposites (PU-Ag) with different concentrations of silver nanoparticles.

Table 1. Keys to identify the nanocomposites of polyurethane and silver and mechanical and thermal properties comparison between PU and the nanocomposites of PU and silver.

Key	Wt% of silver nanoparticles	Storage Modulus E' (MPa)		Td1 (°C)	Td2 (°C)
		−100 °C	37 °C		
PU	0	438	4.9	310	380
PU–1	5.69*10^{-4}	495	3.4	320	380
PU–2	14.20*10^{-4}	566	4.9	320	380
PU–3	28.40*10^{-4}	660	5.6	325	380
PU–4	56.90*10^{-4}	675	6.1	310	370

The degradation temperature of PU increased with the addition of silver nanoparticles to the polymer matrix. In the scientific community, it is believed that small amounts of metallic silver nanoparticles in the polymer matrix can restrict the mobility of polymer chains and act as nucleation sites, encouraging crystallization of nanocomposite materials. The results in Table 1 suggest that small amounts of silver nanoparticles can improve the composite's thermal stability as demonstrated by the degradation temperatures of composite materials which were evaluated by thermal analysis. It should be noted that similar results were observed by Shan-hui Hsu and coworkers with gold nanoparticles in polyurethane composite. The PU material showed an improvement in the mechanical and thermal properties with the addition of gold nanoparticles [17-19]. Similar results were obtained in a polyurethane composite synthesized by suspension and reinforced with silver nanoparticles with an average particle size of 4-7 nm [20].

Table 1 also demonstrates that the decomposition temperature of the soft and rigid blocks of composite materials has increased from 310 °C, for PU without silver nanoparticles, to 325 °C for the polyurethane nanocomposite with 0.0057 wt% of silver nanoparticles. Though it should be noted higher concentrations of silver nanoparticles triggered a decrease in thermal stability. This change in the thermal behavior in the polymer composites may be attributed to the silver nanoparticles cluster formation.

Figure 1a shows the thermomechanical spectrum of the PU without silver nanoparticles and PU composite materials (PU-Ag). This Figure demonstrates that the addition of silver nanoparticles into the PU polymer matrix does not alter the materials' glass transition temperature (Tg). However, at lower temperatures than Tg, the storage modulus (E') raises with the increase of silver nanoparticles content (Figure 1b). This result may be related to the induced crystallinity in polyurethane soft segments due to the presence of silver nanoparticles. Previous experiments have demonstrated that the interaction in the nanoparticle-polymer interface is related to the E' behavior in the dynamic mechanic analysis, as shown in the case of nanocomposite materials with polymer matrices [18].

In general, the addition of inorganic nanoparticles, generally causes a lack of movement in polymer chains, resulting in the increased stiffness of the composite materials [19], which is reflected in E' at -100 °C and in general at 37 °C (Table 1).

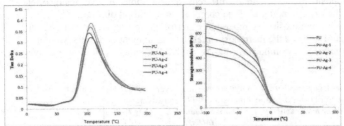

Figure 1. a) Tan delta of the PU without silver nanoparticles and composites materials of PU-Ag and b) Storage modulus of the PU without silver nanoparticles and the series of PU-Ag nanocomposite materials.

Analysis of the dispersion of silver nanoparticles in the polyurethane matrix

Figure 2 shows the surface morphology of the PU without silver nanoparticles and the series of polyurethane composite materials with different concentrations of silver nanoparticles. A micrograph of the PU shows a rough surface, which may be the result of microsegregation that occurs in the system due to the different nature of the segments composing the material. In contrast, PU-Ag-1, which contains 0.00057 wt% of silver nanoparticles, shows a smoother surface. This may be the result of a better interaction between the PU segments promoted by the silver nanoparticle present in the polymer matrix. Since the phase segregation is strongly influenced by the hydrogen interactions [16, 17, 19], it should be considered that a certain concentration of silver nanoparticles can improve the hydrogen bridge formation. This can be studied through FT-IR, the comparison of the absorbance of free and associated carbonyl group in the rigid PU segment. This carbonyl group shows a characteristic absorption band depending on whether it is free or associated.

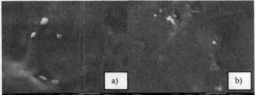

Figure 2. (a) Micrograph of the PU without silver nanoparticles and (b) nanocomposite material PU-Ag-1

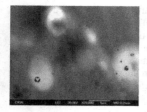

Figure 3. Micrograph of polyurethane nanocomposite PU-Ag-4

When the silver nanoparticles concentration exceeds a certain concentration, the formation of nanoparticle clusters commences, and the process then breaks apart the hydrogen bridges between the segments of PU seen by FT-IR (not shown here). Similar results have been reported by Chou, *et. al.* [20] in PU matrices reinforced with silver nanoparticles. Figure 3 is a micrograph of PU-Ag-4, one of the nanocomposites with a higher concentration of nanoparticles. In the micrograph, dark spots can be observed, and are probably due to the nanoparticles clusters which are responsible for the breaking of hydrogen bonds in the PU matrix and for altering the surface characteristics and properties of the samples [21, 22].

Antifungal properties of polyurethane nanocomposites

In Figure 4 petri dishes containing growth media and the composite materials can be seen, along with the *T. mentagrophytes* fungus that was grown in the dishes. After 168 hours of incubation with adequate conditions, the samples were observed. The PU without silver nanoparticles with the lack of a growth inhibition halo can be seen in image 4a), indicating that the material has no antifungal properties.

Figure 4. Photograph of a) The PU without silver nanoparticles, b) PU-Ag-3 with 0.00284 wt% of silver nanoparticles and c) PU-Ag-4 with 0.0057 wt% of silver nanoparticles.

In contrast, a growth inhibition halo against the fungus is clearly observed in image 4b, corresponding to the polyurethane nanocomposite of PU-Ag-3. There is no growth of *T. mentagrophytes* vicinity of the composite material, indicating the presence of antifungal activity provided by the silver nanoparticles embedded within the polymer matrix. The results show that the silver nanoparticles have a considerable biocide activity [23]. Interestingly, increasing the silver nanoparticles concentration in the composite material (Figure 4, image c) resulted in the loss of antifungal capacity as illustrated by the absence of a halo of inhibition near the composite material. Many studies have shown the antimicrobial effects of silver nanoparticles, but the effectiveness of the nanoparticles against fungal pathogens of the skin including clinical isolates of *T. mentagrophytes* are mostly unknown.

Presumably the behavior of the antifungal activity of the silver nanoparticles is also related to the appearance of clusters. It is likely those clusters decrease the ability of nanoparticles to inhibit the fungus growth, as could be seen in the physical-mechanical properties of the composite materials where it was observed that increasing the silver nanoparticles concentration decreases some properties like the degradation temperature. This aggregates formation has been documented by Sanchez, et. al [24] where they observed an increase in the silver nanoparticles aggregation as the concentration increases in composites of polyethylene/silver nanoparticles.

CONCLUSIONS

The results show that the addition of small amounts of silver nanoparticles may improve the thermal and mechanical performance of the polyurethane composite materials by the formation of hydrogen bridges. The hydrogen bridges then improve the thermal stability of the materials, followed by the phase interactions and the increased mechanical performance. However, it is important to note that with a higher concentration of silver nanoparticles there is a decrease in the antifungal performance.

REFERENCES

1. Han DK, Park K, Park DD, Ahn KD, and Kim YH (2006), *Artif. Organs* 30: 955-9.
2. Williams RL, Krishna Y, Dixon S, Haridas A, Grieson I and Sheridan C (2005), *J., Mater. Sci.* 16:1087-92
3. Santerre JP, Woodhouse K, Laroche G, and Labow RS (2005). *Biomaterials* **26:**7457-70.
4. Makosco CW (1989) RIM Fundamentals of Reaction Injection Molding, Hanser Publishers, New York.
5. Gersappe D (2002), *Phys. Rev. Lett.*, **89** 058301.
6. Mitra SB, Wu D and Holmes BN (2003), *J., Am., Dent. Assoc.*, **134:**1382-90.
7. Balazs AC, Emrick T and Russel TP (2006). *Science*, **314:1**107-10.
8. Klasen HJ (2000), Burns 26:131-138.
9. Pollini M, Russo M, Licciulli A, *et al* (2009), *J. Mat. Sci. Mat. in Medicine*, **20,** 11:2361-2366
10. Edwards-Jones V (2009), *Letters in applied microbiology*, **49**, 2:147-152.
11. Sang Young Yeo, Hoon Joo Lee, Sung Hoon Jeong (2003), *J. Mat. Sci.*, **38:**2143-2147.
12. Matsumura Y, Yoshikata K, Kunisaki S, Tsuchido T (2003), *Appl. Environ. Microbiol*, **69**:4278-81.
13. Feng QL, Wu J, Chen GQ, Cui FZ, Kim TN, Kim JO (2000), *J. Bio. Mat. 5, 4:* 662-668.
14. Gupta A, Maynes M, Silver S (1998), *Appl. Environ. Microbiol.* **64:**5042-5045.
15. Elechiguerra JL *et al* (2005), *J. Nanobiotech.*, **3**:6
16. Keuk-Jun Kim, Woo Sang Sung, Seok-Ki Moon, Jong Guk Kim, (2008), *J. Microbiol. Biotechnol.* **18**, 8:1482-1484.
17. Shan-Hui H, Chih-Wei C and Sheng-Mo T (2004), *Macromol. Mat. Eng.* **289**, 12:1096-1101.
18. Landry CJT, Coltrain BK, Landry MR, Fitzgerald JJ, Long VK (1993) *Macromolecules*, **6**:3702.
19. Huang HH, Wilkes GL, Carlson JG (1989) *Polymer*, **30:**2001-2012.

20. Chou CW, Hse SH, Chang H, Tseng SM, Lin HR (2005), *Polymer Degradation and Stability*, **91:**1017-24.
21. Senich GA, MacKnight WJ (1980) *Macromolecules*, 13:106.
22. Sung CSP, Schneider NS (1978) *J. Mater. Sci.*, 13:1689.
23. Kim, Keuk-Jun, Woo Sang Sung, Seok-Ki Moon, Jong-Soo Choi, Jung-Guk Kim and Dong-gun Lee (2008), *J. Microbiol. Biotechnol.* **18**, 8:1482-1484.
24. Sanchez Valdez S, Ortega Ortíz H, Ramos de Valle L F, Medellin Rodríguez F J and Guedea Miranda R, (2009), *Journal Applied Polymer Science*, 111: 953-962.

Mater. Res. Soc. Symp. Proc. Vol. 1547 © 2013 Materials Research Society
DOI: 10.1557/opl.2013.684

Innovative gold nanoparticle patterning and selective metallization

E.S. Kooij[1], M.A. Raza[1,2], H.J.W. Zandvliet[1]

[1]Physics of Interfaces and Nanomaterials, MESA+ Institute for Nanotechnology, University of Twente, P.O. Box 217, 7500AE Enschede, The Netherlands

[2]Centre of Excellence in Solid State Physics, University of the Punjab, QAC, Lahore-54590, Pakistan

ABSTRACT

We present a simple, novel procedure to selectively deposit gold nanoparticles using pure water. It enables patterning of nanoparticle monolayers with a remarkably high degree of selectivity on flat as well as microstructured oxide surfaces. We demonstrate that water molecules form a thin 'capping' layer on exposed thiol molecules within the mercaptan self-assembled layer. This reversible capping of water molecules locally 'deactivates' the thiol groups, therewith inhibiting the binding of metallic gold nanoparticles to these specific areas. In addition, we show that this amazing role of water molecules can be used to selectively metalize the patterned gold nanoparticle arrays. Employing an electroless seeded growth process, the isolated seeds are enlarged past the percolation threshold to deposit conducting metal layers.

INTRODUCTION

In the fabrication technology based on implementing functional nanoparticles, one of the key challenges is the patterned deposition of the nanoentities on pre-defined areas by simple, fast and low-cost methods. Various 'top-down' and 'bottom-up' techniques with sophisticated equipment and chemicals have been used to achieve the goal of nanoscale patterning. The majority of nano- and micropatterning techniques are in the former category, including photolithography, electron-beam lithography, microcontact printing, dip-pen nanolithography and laser-based patterning. Despite its enormous impact on modern technology, the top-down approach has a number of disadvantages and limitations. For example photolithography and electron-beam lithography techniques require hazardous chemicals. This in turn restricts their use in patterning nanoparticles or molecules with organic functionalities, because these chemicals may destroy the organic molecules and biological entities. Microcontact printing requires an elastomeric stamp to fabricate heterogeneous structures of typically micrometer dimensions, but use for a mixed functionality surface fabrication is limited [1]. In principle, dip-pen nanolithography enables nanoscale patterning [2]. However, due to the relativity low transfer efficiency the limited amount of materials transported to the substrate hinders large scale fabrication. Similarly, the laser-based patterning techniques such as laser-induced photothermal deposition [3] and laser-based particle deposition [4] can be used to pattern nanoparticles and nanoclusters, but again these technologies demand a sophisticated laser system and a complicated deposition process involving many steps.

The alternative 'bottom-up' approaches resolve a number of drawbacks and limitations of the aforementioned methods. Moreover, various combinations of both 'top-down' and 'bottom-up' have also been suggested. An example of such a combined approach is referred to as 'microcontact deprinting' [5] enabling hierarchical patterning of nanoparticles on a wide range of substrates. Although this is fast and versatile, it requires a plasma system to burn off the polymer micelles together with complicated procedures to apply and peel off the polymer stamp at high temperature therewith increasing the overall complexity.

Here we describe a simple, fast and low-cost method to enable patterned deposition of nanoparticles and demonstrate local metallization. Previously we attempted to locally derivatize the surface with amino- and mercaptosilane molecules by microcontact printing [6]. Although there was an apparent density difference, the selectively was limited. The present approach does not require sophisticated equipment nor does it involve harsh chemical procedures.

LOCAL NANOPARTICLE DEPOSITION

For specific experimental details, we refer to our recent work [7]. In figure 1(a) our novel method to pattern gold nanoparticle deposits is schematically shown. In the first step a cleaned Si/SiO$_2$ substrate is derivatized with mercaptosilane (MPTMS). Subsequently, a specific area of the substrate is treated by bringing it into contact with pure water. Finally, gold nanoparticles are deposited by immersing the substrate into a nanocolloidal gold solution. In figure 1(b) a photograph of the sample after water treatment using a mm-sized droplet and subsequent gold nanoparticle deposition is shown. Clearly a circular spot can be identified where the water has been in contact with the MPTMS layer. A clear difference in reflective properties, i.e. contrast can be discerned between the areas of the surface which have been water-treated (WT) and the areas which have not been in contact with water (non-WT). We assume that the difference in contract observed in the macroscopic images arises from the fact that gold nanoparticles could not be attached to the WT area, while the untreated surface is fully covered with nanoparticles.

To verify this assumption, in figure 1(b,right) a Helium ion microscopy (HIM) image is shown of silica coated with 50nm gold nanoparticles, in the border area between non-WT and WT areas. The untreated area (right part) shows a dense coverage of nanoparticles indicating the irreversible deposition of these nanoentities as described previously [8]. The strong interaction between the exposed thiol groups of the MPTMS molecules induces a strong affinity to irreversibly bind the gold nanoparticles to the surface. Surprisingly, on the area within the circle where the MPTMS was in contact with the water droplet, gold nanoparticles do not adsorb. Remarkably, we did not even find a single gold particle, revealing the high degree of selectivity.

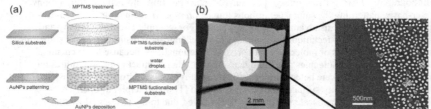

Figure 1. (a) Schematic representation of the nanoparticle patterning procedure using water. After derivatization of the surface with MPTMS, selective parts are defunctionalized by water treatment, followed by site-selective deposition of gold nanoparticles. (b) Macroscopic view and microscopic images of gold nanoparticle patterning on a flat Si/SiO$_2$ substrate. The left and right images, respectively, are a photograph after patterned gold nanoparticle deposition and a Helium ion microscopy image representing the indicated area.

PATTERNED FUNCTIONALIZATION

To address the effect of water on the MPTMS-coated area, different possible scenarios can be envisaged. A first assumption involves removal of the layer of MPTMS from the WT area by washing off by the water droplet. This would explain why gold nanoparticles are not adsorbed in the WT areas since removal of MPTMS also leads to the absence of free thiol groups available for gold binding. A second possibility is that by water treatment the free thiol end-groups are oxidized to form sulfate ions. Therewith the functionality to irreversibly bind gold nanoparticles is lost. It is well-documented in literature that the presence of physisorbed water and oxygen at the MPTMS surface may lead to the oxidation of mercaptan [9–11]. In another possible oxidation process the thiol end groups are transformed into disulfide S-S entities with a low reactivity [12]. The replacement of thiol by disulfide groups could also account for the strongly reduced affinity to irreversibly bind gold nanoparticles.

In view of the aforementioned possible accounts for the effect of water on MPTMS monolayers, and to elucidate the potential mechanism, we performed an essential experiment, which in fact rules out any of the aforementioned models. As described above, after water treatment the affinity to bind gold nanoparticles is completely destroyed. However, when samples are heated to approximately 120°C for a short period of time, typically 20 minutes, the functionality of the thiol molecules appears to be fully restored. As with non-WT samples, gold nanoparticle deposition is again possible after heating WT samples. In all cases, we did not observe any notable difference in deposition characteristics between situations (i) before water treatment and (ii) after water treatment and subsequent heating. This indicates the complete reversibility of the water treatment by simply heating the samples. None of the models described above can account for this observation. The washed-off MPTMS molecules are not 'restored' during heating, while possibly oxidized states (either sulfate or disulfide entities) will not be restored to their reduced states simply by heating in an oxidizing environment.

We present an alternative explanation for the reversible disappearance of the gold binding functionality of the thiol groups. When considering the electronegativities of oxygen, sulfur and hydrogen (3.44, 2.58 and 2.2, respectively) [13], the differences between values for the constituting elements in water (1.24) and thiol entities (0.38) indicate that both the OH and the SH bonds are polarized. As such, there will be an attractive electrostatic interaction between the water on one side and the thiol entities on the other. As such, we envisage that a bonding between the OH and SH bonds occurs (see figure 2), giving rise to hydrogen-bond type binding of a water monolayer to the MPTMS self-assembled monolayer. Owing to the aforementioned differences in electronegativity of oxygen and sulfur atoms, the square-like hydrogen bonding will most likely not be as symmetric as schematically shown in figure 2(a).

The possibility for thiol end-groups to form hydrogen bonds has been discussed in a number of literature reports [14,15]. Using infrared spectroscopy, possible signatures of thiol-based hydrogen bonding have been observed and discussed [16,17]. Evidence for weak hydrogen bonds in thiols has also been derived from magnetic resonance experiments [18,19]. Colebrook and Tarbell state that previous literature reviews indicate that thiol-based hydrogen bonds can be formed with strong donor groups such as oxygen. But they also suggest the possibility that the sulfur atom can act as a donor in hydrogen bonding [18]. Finally, in a more recent paper Yadav *et al.* [20] discuss the possible effect of water on the Michael addition. The proposed mechanism involves a double hydrogen bond between the water molecule and quinone compounds. The 'capping' of water molecules produces a thin layer of water on the thiol entities, which inhibits the effective binding of gold nanoparticles to the available sulfur groups on the surface. Simply said, the water molecules effectively act as a 'resist' enabling patterning of whatever binds to the thiol-terminated MPTMS molecules.

This tentative mechanism would explain why gold nanoparticles are absent on water-treated regions. Also, it accounts for the restoration of the affinity to bind gold simply by heating the samples. Nevertheless, a surprising aspect of this tentative model is that in the case of non-WT samples, which exhibit a high affinity for gold, the nanoparticles are deposited from aqueous solutions. Why do gold particles not bind to previously water treated MPTMS layers, while they do adsorb in the presence of water?

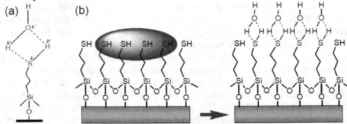

Figure 2. Schematic illustration (a) of the water molecule binding to the MPTMS molecule due to the electronegativity difference between O, S and H, and (b) the capping of water molecules 'shielding' exposed thiol molecules, therewith hindering gold-sulfur bonding after water treatment on MPTMS and thus inhibiting gold nanoparticle deposition.

SELECTIVE METALLIZATION

During selective metallization, the patterned gold nanoparticle arrays described in the previous section act as seed particles to the electroless gold plating. A previously reported method was employed to grow continuous metal layers by electroless deposition [21,22]. The growth solution essential consists only of the precursor HAuCl$_4$ and NH$_2$OH as reducing agent. Typically within 20-30 minutes at room temperature, a continuous gold layer is formed specifically on the prepatterned areas, as is shown in figure 3. The surface areas which have been in contact with water can clearly be identified as circular spots on the sample after gold nanoparticle deposition and subsequent metallization.

The morphology of pre-patterned samples after seeded gold growth reveals a remarkable contrast between WT and non-WT areas. The boundary between the different regions appears to be very sharp (electron microscopy image of figure 3), clearly distinguishing between areas where gold nanoparticles were present and where they were absent. Apparently, the gold metallization is also highly selective, leading to a thick gold film only in the non-WT sections. The gold nanoparticles act as catalysts for the electroless plating to form a dense metallic layer. The morphology of the thick metallic layer in the non-WT area, i.e. outside the circular spot, is characterized by a large number of densely packed grains of different sizes.

Owing to the absence of nanoparticles in the WT area such a metallic layer could not develop. However, after prolonged deposition, typically up to 30-40 minutes, randomly distributed islands of different sizes in the low micrometer range develop within the WT circular spot (not shown in figure 3). Our initial assumption was that these relatively large clusters form in solution and subsequently sediment onto the surface under the influence of gravity. However, changing the orientation of the substrate during seeded growth, even turning it upside down, reveals that the clusters are still formed in the WT regions. So apparently, these islands somehow nucleate within the defunctionalized regions, and grown into large entities with prolonged growth.

Further characterization and analysis of the electroless growth of gold on the patterned seed particle arrays reveals that the growth process is similar to what has been reported for silver growth [6,23]. During the first few minutes, the electroless deposition gives rise to enlargement of the individual seed particles into large, but still isolated islands. After approximately 10 minutes growth a percolating network of interconnected gold islands is formed, as is also confirmed by the fact that a finite resistivity can be measured. Prolonged seeded growth eventually gives rise to a fully developed continuous layer, as is shown in figure 3(right). For films with a thickness up to a micrometer, the conductivity was found to saturate at approximately 30% of the bulk conductivity.

Figure 3. Optical image (left) and enlarged electron microscopy (right) image after 15 minutes electroless metallization following water-based patterning. The circular areas were in contact with the water. Subsequent selective seeded growth gives rise to continuous metal layers in the non-WT areas. In the electron microscopy image, the granular structure of the metallic layer can be discerned.

CONCLUSIONS

We have shown that pure water can be used to selectively 'defunctionalize' MPTMS coated surfaces. We propose that water molecules form a thin 'capping' layer on exposed thiol molecules within the mercaptan self-assembled layer. This reversible capping of water molecules locally 'deactivates' the thiol groups, therewith inhibiting the binding of metallic gold nanoparticles to these specific areas. We have demonstrated the remarkably high degree of selectivity in the self-assembled formation of these nanogold monolayers from solution. Since MPTMS is used as a binding agent not only for gold but also for other materials, the method described here can in principle be extended to enable patterning of other nanoscale materials.

In addition to the localized gold nanoparticle deposition, we studied the selective metallization by electroless gold deposition on the patterned seed particles. The latter process gives rise to enlargement of the isolated seeds past the percolation threshold, to ultimately give rise to conducting metal structures. The obtained patterned gold films exhibit macroscopic conductivity values approximately a factor of three lower than that of bulk gold.

ACKNOWLEDGMENTS

The authors thank Dr. G. Hlawacek (University of Twente) for performing the HIM measurements and for helpful discussion. We gratefully acknowledge the support by MicroNed, a consortium to nurture micro-systems technology in The Netherlands. One of the authors (M.A. Raza) acknowledges support from the Higher Education Commission in Pakistan.

REFERENCES

Reprinted from *J. Colloid Interface Sci.* **364**, M.A. Raza, E.S. Kooij, A. van Silfhout, H.J.W. Zandvliet, B. Poelsema, *Novel, highly selective gold nanoparticle patterning on surfaces using pure water*, pp. 304-310. Copyright (2011), with permission from Elsevier.

1. L. Libioulle, A. Bietsch, H. Schmid, B. Michel, E. Delamarche, *Langmuir* **15**, 300 (1999).
2. R.D. Piner, J. Zhe, F. Xu, S. Hong, C.A. Mirkin, *Science* **283**, 661 (1999).
3. L. Scheres, B. Klingebiel, J. ter Maat, M. Giesbers, H. de Jong, N. Hartmann, H. Zuilhof, *Small* **6**, 1918 (2010).
4. J. Xu, J. Drelich, E.M. Nadgorny, *Langmuir* **20**, 1021 (2004).
5. J. Chen, P. Mela, M. Möller, M.C. Lensen, *ACS Nano* **3**, 1451 (2009).
6. A.A. Mewe, E.S. Kooij, B. Poelsema, *Langmuir* **22**, 5584 (2006).
7. M.A. Raza, E.S. Kooij, A. van Silfhout, H.J.W. Zandvliet, B. Poelsema, *J. Colloid Interface Sci.* **364**, 304 (2011).
8. E.S. Kooij, E.A.M. Brouwer, H. Wormeester, B. Poelsema, *Langmuir* **18**, 7677 (2002).
9. J. Singh, J.E. Whitten, *J. Phys. Chem. C* **112**, 19088 (2008).
10. J.J. Senkevich, G.R. Yang, T.M. Lu, *Colloids Surf. A* **207**, 139 (2002).
11. S.R. Yang, B.O. Kolbesen, *Appl. Surf. Sci.* **255**, 1726 (2008).
12. G. Ledung, M. Bergkvist, A.P. Quist, U. Gelius, J. Carlsson, S. Oscarsson, *Langmuir* **17**, 6056 (2001).
13. D.R. Lide (ed.), *CRC Handbook of Chemistry and Physics*, Taylor & Francis, Boca Raton, Fl. (2005).
14. I.V. Zuika, Y.A. Bankovskii, *Russ. Chem. Rev.* **42**, 22 (1973).
15. F.H. Allen, C.M. Bird, R.S. Rowland, P.R. Raithby, *Acta Cryst. B* **53**, 696 (1997).
16. A. Menefee, D. Alford, C.B. Scott, *J. Chem. Phys.* **25**, 370 (1956).
17. O.P. Yablonskii, N.M. Rodionova, L.F. Lapuka, *J. Appl. Spectrosc.* **19**, 1303 (1973).
18. L.D. Colebrook, D.S. Tarbell, *Proc. Natl. Acad. Sci. USA* **47**, 993 (1961).
19. S.H. Marcus, S.I. Miller, *J. Am. Chem. Soc.* **88**, 3719 (1966).
20. J.S. Yadav, T. Swamy, B.V.S. Reddy, D.K. Rao, *J. Mol. Catal. A: Chem.* **274**, 116 (2007).
21. K.R. Brown, L.A. Lyon, A.P. Fox, B.D. Reiss, M.J. Natan, *Chem. Mater.* **12**, 314 (2000).
22. S.M. Tabakman, Z. Chen, H.S. Casalongue, H. Wang, H. Dai, *Small* **7**, 499 (2011).
23. A.J. de Vries, E.S. Kooij, H. Wormeester, A.A. Mewe, B. Poelsema, *J. Appl. Phys.* **101**, 053703 (2007).

Mater. Res. Soc. Symp. Proc. Vol. 1547 © 2013 Materials Research Society
DOI: 10.1557/opl.2013.541

A Thermal Decomposition Approach for the Synthesis of Iron Oxide Microspheres

Geetu Sharma[1] and Jeevanandam Pethaiyan[1]

[1]Department of Chemistry, Indian Institute of Technology Roorkee, Roorkee-247667, India

ABSTRACT

Iron oxide microspheres possess a wide range of applications in lithium storage batteries, sensors, photocatalysis, environmental remediation, magnetic resonance imaging and drug delivery. The most commonly used method for the preparation of iron oxide microspheres is hydrothermal synthesis. Besides this, other synthetic methods such as co-precipitation, electrostatic self- assembly, microwave and sol-gel have been reported. The reported synthetic methods usually require longer time (2 to 48 hours) and expensive experimental set up. In the present study, a novel low temperature thermal decomposition approach for the synthesis of iron oxide microspheres has been reported. Thermal decomposition of an iron-urea complex ($[Fe(CON_2H_4)_6](NO_3)_3$) in a mixture of diphenyl ether and dimethyl formamide at 200 °C for 35 minutes leads to the formation of iron oxide microspheres. The microspheres were characterized using a variety of analytical techniques such as X-ray diffraction (XRD), field emission scanning electron microscopy (FE-SEM), transmission electron microscopy (TEM), diffuse reflectance spectroscopy (DRS) and magnetometry. The XRD results indicated amorphous nature for the as prepared iron oxide, whereas after calcination at 500 °C, crystalline α-Fe_2O_3 phase is obtained. The SEM images indicated uniform spheres with an average diameter of 1.2 ± 0.3 μm. The DRS results too gave evidence for the formation of α-Fe_2O_3 on calcination of the microspheres at 500 °C. The field and temperature dependent magnetic measurement results indicated superparamagnetic behavior for the as prepared iron oxide microspheres indicating that the microspheres consist of iron oxide nanoparticles. On the other hand, an antiferromagnetic behavior was observed for the microspheres calcined at 500 °C. The present synthetic method is a novel method to produce magnetic materials with controlled morphologies.

INTRODUCTION

Iron oxide nanoparticles have been used in a wide range of applications such as energy storage, catalysis, sensors, magnetic resonance imaging and drug delivery [1]. These applications are based on the unique magnetic properties of iron oxide particles which vary with their size and shape [2]. Synthesis of iron oxide particles with specific size and well defined morphology has been extensively investigated. Various morphologies such as sphere, star and cube [3], octahedron [4], worm [5], diamond, prism and hexagon [6] and hollow sphere [7] have been reported. To make effective use of iron oxide, recently much attention has been directed towards the synthesis of hierarchical iron oxide nanostructures formed by self-assembly such as flower-like, dendrite-like, chain and rod-like particles and microspheres [8-11].

Recently, iron oxide microspheres are of immense interest because of their widespread applications in lithium ion storage batteries [11], sensors [12], water treatment [13], catalysis [14] and imaging [15]. Xiong et al. have reported superior performance of iron oxide microspheres in lithium ion batteries compared to other nanostructured materials [11]. Song et al. have reported enhanced sensing response using α-Fe$_2$O$_3$ hollow microspheres compared to pure α-Fe$_2$O$_3$ nanoparticles [12]. Liu et al. have reported better photocatalytic activity using α-Fe$_2$O$_3$ microspheres compared to nano and micron-sized α-Fe$_2$O$_3$ particles [14]. The iron oxide microspheres have been assembled from different building blocks such as nanorods [12], nanoparticles [16], nanosheets [17], tetrahedron [18] and octahedron units [19]. Various synthetic methods have been reported for the preparation of iron oxide microspheres. The most commonly used method is hydrothermal synthesis [12, 16, 18]. Besides this, other methods such as polymerization induced colloidal aggregation [20], microwave-solvothermal [21], electrostatic self-assembly [22], sol-gel [23] and template method [24] have been employed. The reported methods often require longer reaction times (e.g. 2 to 48 h), special apparatus (e.g. autoclave and microwave), inert conditions and sometimes involve complicated reaction steps. Recently, prismatic iron oxide particles formed by self-assembly have been synthesized using a simple thermal decomposition approach [25]. In the present study, a one step synthesis of iron oxide microspheres using the thermal decomposition of iron-urea complex, in a mixture of dimethyl formamide and diphenyl ether, is reported.

EXPERIMENTAL DETAILS

Ferric nitrate nonahydrate was received from SD Fine Chemicals, urea and dimethyl formamide (DMF) were received from Rankem, and diphenyl ether (DPE) was received from Sigma-Aldrich and the reagents were used as received. The iron-urea complex, Fe[CON$_2$H$_4$]$_6$(NO$_3$)$_3$, was prepared by a previously reported method [26]. The procedure for the synthesis of iron oxide microspheres is as follows. About 500 mg of iron-urea complex was dissolved in 7 mL of dimethyl formamide in a round bottom flask. To the obtained yellow colored solution, 3 mL of diphenyl ether was added. The contents were refluxed at 200 °C in air for a period of 35 minutes with continuous stirring. After the completion of reaction, the contents were naturally cooled to ambient temperature and about 30 mL of methanol was added. The precipitate obtained was centrifuged, washed repeatedly with methanol and kept for drying in air. The brown colored product was calcined in air at 500 °C at a heating rate of 1 °C/min inside a muffle furnance (Nabertherm®).

RESULTS AND DISCUSSION

Fig. 1a shows the powder XRD patterns of iron oxide samples before and after calcination. The as prepared sample is amorphous to X-rays. This indicates that the iron oxide particles are so small that no detectable diffraction can be observed. The decomposition of iron-urea complex in air under refluxing conditions (i.e. heating the contents close to the boiling pointing of the solvent and keeping at this temperature with a reflux condenser) provides an oxidation atmosphere. Under these conditions, γ-Fe$_2$O$_3$ is formed [27]. After calcination at 500 °C, the sample shows diffraction peaks due to α-Fe$_2$O$_3$ (JCPDS file no. 85-0987). The presence of γ-Fe$_2$O$_3$ and α-Fe$_2$O$_3$ phases in the as prepared and the calcined samples, respectively, is further evidenced from magnetic measurement results discussed later.

The SEM image of as prepared iron oxide (Fig. 1b) shows the microspheres with a narrow size distribution (mean diameter = 1.2 ± 0.3 µm). A representative TEM image of the same sample (Fig. 1c) shows well developed iron oxide microspheres. The selected area electron

156

diffraction pattern (inset in Fig. 1c) shows the amorphous nature of as prepared iron oxide microspheres, which is in accordance with the XRD results.

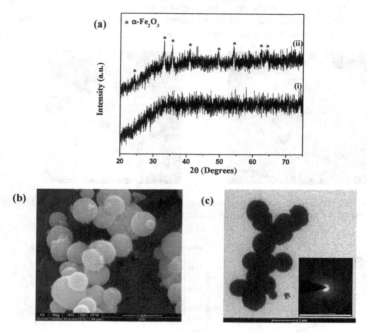

Fig. 1: (a) Powder XRD patterns of iron oxide; (i) as prepared, and (ii) after calcination at 500°C. (b) FE-SEM image and (c) TEM image of as prepared iron oxide. The inset in Fig. 1c shows the SAED pattern.

To understand the effect of heat treatment on the morphology of the spheres, calcination of the as prepared iron oxide was carried out at 500 °C and the corresponding SEM and TEM images are shown in Fig. 2. It can be noticed that the microspheres retain their morphology after calcination at 500 °C. The bright field TEM image of iron oxide after calcination shows that the microspheres are further made up of small particles (Fig. 2b). The SAED pattern (inset in Fig. 2b) indicates polycrystalline nature of iron oxide microspheres after calcination.

Fig. 3 shows the zero field cooled (ZFC) and field cooled (FC) magnetization curves as a function of temperature for the as prepared and calcined iron oxide microspheres measured under an applied field of 100 Oe. The ZFC curve for the as prepared iron oxide microspheres (Fig. 3a) shows a sharp maximum at 48 K indicating the blocking transition of superparamagnetic iron oxide nanoparticles. The FC curve for the as prepared iron oxide microspheres shows a slight decrease below the blocking temperature. This characteristic behavior is attributed to strong interparticle interactions in γ-Fe_2O_3 nanoparticles that make up the microspheres [28]. From the observed T_B, the average particle size for γ-Fe_2O_3 nanoparticles was calculated as

19 nm. After calcination at 500 °C, the iron oxide microspheres show the characteristic Morin transition at 255 and 225 K in the ZFC and FC curves, respectively, indicating the formation of α-Fe$_2$O$_3$ (Fig. 3b). Below the Morin transition, the magnetic sub-lattices in α-Fe$_2$O$_3$ are antiparallel and the material behaves as an antiferromagnet. Above the Morin transition, the material exhibits weak ferromagnetism due to spin canting [29].

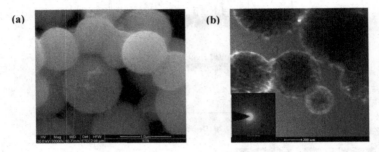

Fig. 2: (a) FE-SEM image of iron oxide microspheres after calcination at 500 °C and (b) bright field TEM image of the same sample (the inset shows the SAED pattern).

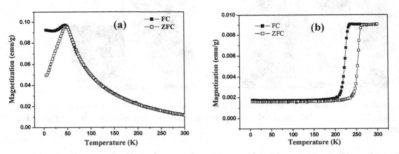

Fig. 3: ZFC-FC curves of iron oxide microspheres: (a) as prepared, and (b) after calcination at 500 °C.

In the thermal analysis studies, the as prepared iron oxide microspheres showed a total weight loss of 28 % (up to about 400°C) indicating the presence of organic moieties on the surface of iron oxide microspheres. The differential thermal analysis pattern of the as prepared iron oxide showed exothermic peaks at 206 °C, 240 °C, 303 °C and 448 °C. The peaks at 206 °C,   240 °C and 303 °C were attributed to the removal of organic molecules. The peak at 448 °C was attributed to the phase transformation of γ-Fe$_2$O$_3$ to α-Fe$_2$O$_3$. Elemental analysis on the as prepared iron oxide microspheres indicated the presence of C (4.8%), N (3.6%) and H (0.7%). FT-IR spectra of as prepared and calcined iron oxide samples showed bands near 3400 cm^{-1} (O-H /N-H stretching) and 1630 cm^{-1} (O-H bending). In the as prepared iron oxide, additional bands were observed at 1469 cm^{-1}, 1370 cm^{-1} and 1052 cm^{-1} that were attributed to C-N stretching, and bending vibrations of carbonyl and N–H groups, respectively. The bands at 590 cm^{-1} and 480 cm^{-1} in the as prepared iron oxide were attributed to γ-Fe$_2$O$_3$. After calcination at 500 $^\circ$C, characteristic bands due to α-Fe$_2$O$_3$ at about 550 cm^{-1} and 469 cm^{-1} were observed. The IR bands due to organic moieties disappeared on calcination at 500 $^\circ$C indicating that the organic moieties are lost on calcination.

Based on the TG-DTA, FT-IR and elemental analysis results, a possible mechanism for the formation of iron oxide microspheres by the thermal decomposition of iron-urea complex is depicted in Fig. 4. It is proposed that cyanuric acid is formed along with iron oxide nanoparticles on the thermal decomposition of iron-urea complex [25]. Iron-urea complex decomposes with melting at about 178 $^\circ$C to ferric nitrate and urea. Urea converts to isocyanuric acid at 140 $^\circ$C and ferric nitrate decomposes at about 200 $^\circ$C to form iron oxide. Isocyanuric acid on further reaction with urea at 190 $^\circ$C forms biuret which converts finally to cyanuric acid. The cyanuric acid molecules present on the surface of iron oxide nanoparticles are hydrogen bonded to dimethyl formamide molecules; DMF was used as one of the solvents in the mixture during the synthesis. The iron oxide nanoparticles coated with the organic molecules self-assemble to form the iron oxide microspheres. The non-selective adsorption of ligand molecules on the surface of iron oxide leads to isotropic growth of particles which in turn self-assemble to form the microspheres.

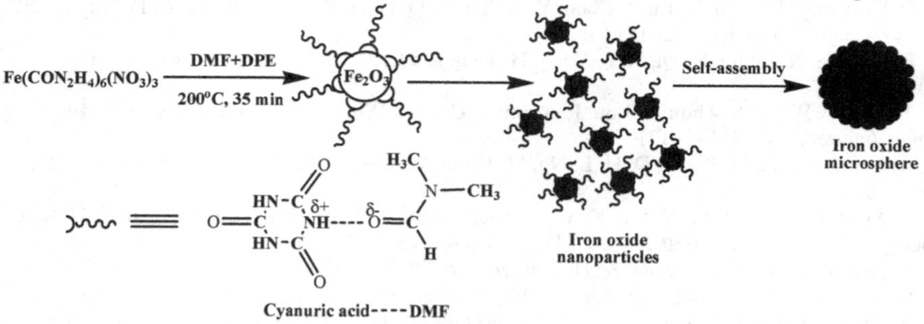

Fig. 4: Mechanism of formation of iron oxide microspheres.

CONCLUSIONS

Iron oxide microspheres were successfully synthesized using a novel low temperature thermal decomposition approach. The method is simple, novel and can produce iron oxide particles with controlled morphology. Electron microscopic studies (TEM and SEM) indicate that the microspheres consist of self-assembled iron oxide nanoparticles. Magnetic studies indicate superparamagnetic behavior for the iron oxide microspheres which suggests that the microspheres are made up of self-assembled iron oxide nanoparticles. The iron oxide microspheres are expected to be useful in applications such as lithium-ion storage batteries, sensors, water treatment, catalysis and imaging.

ACKNOWLEDGEMENT

Financial support from Council of Scientific and Industrial Research, Government of India (Project No: 01 (2311) /09/EMR-II) is gratefully acknowledged with gratitude.

REFERENCES
1. C. Yang, J. Wu, and Y. Hou, *Chem. Commun.* **47**, 5130 (2011).
2. X. Mou, X. Wei, Y. Li and W. Shen, *CrystEngComm* **14**, 5107 (2012).
3. L. M. Bronstein, J. E. Atkinson, A. G. Malyutin, F. Kidwai, B. D. Stein, D. G. Morgan, J. M. Perry and J. A. Karty, *Langmuir* **27**, 3044 (2011).
4. L. Zhang, J. Wu, H. Liao, Y. Hou and S. Gao, *Chem. Commun.* 4378 (2009).
5. S. Palchoudhury, Y. Xu, J. Goodwin and Y. Bao, *J. Appl. Phys.* **109**, 07E314 (2011).
6. J. Cheon, N. J. Kang, S. M. Lee, J. H. Lee, J. H. Yoon and S. J. Oh, *J. Am. Chem. Soc.* **126**, 1950 (2004).
7. S. Peng and S. Sun, *Angew. Chem., Int. Ed.* **46**, 4155 (2007).
8. L.Wang, T. Fei, Z. Lou, and T. Zhang, *ACS Appl. Mater. Interfaces* **3**, 4689 (2011).
9. G. Sun, B. Dong, M. Cao, B. Wei, and C. Hu, *Chem. Mater.* **23**, 1587 (2011).
10. M. Muruganandham, R. Amutha, M. Sathish, T. S. Singh, R. P. S. Suri, and M. Sillanpaa, *J. Phys. Chem. C* **115**, 18164 (2011).
11. Q. Q. Xiong, J. P. Tu, Y. Lu, J. Chen, Y. X. Yu, Y. Q. Qiao, X. L. Wang, and C. D. Gu, *J. Phys. Chem. C* **116**, 6495 (2012).
12. H. J. Song, X. H. Jia, H. Qi, X. F. Yang, H. Tang and C. Y. Min, *J. Mater. Chem.* 22, 3508 (2012).
13. Y. Liu, Y. Wang, S. Zhou, S. Lou, L. Yuan, T. Gao, X. Wu, X. Shi and K.Wang, *ACS Appl. Mater. Interfaces* **4**, 4913 (2012).
14. G. Liu, Q. Deng, H. Wang, D. H. L. Ng, M. Kong, W. Cai and G. Wang, *J. Mater. Chem.* **22**, 9704 (2012).
15. S. Xuan, F. Wang, J. M. Y. Lai, K. W. Y. Sham, Y. X. J. Wang, S. F. Lee, J. C. Yu, C. H. K. Cheng, and K. C. F. Leung, *ACS Appl. Mater. Interfaces* **3**, 237 (2011).
16. C. Han, D. Zhao, C. Deng and K. Hu, *Mater. Lett.* **70**, 70 (2012).
17. S. W. Cao, Y. J. Zhu and Y. P. Zeng, *J. Magn. Magn. Mater.* **321**, 3057 (2009).
18. Y. Lv, H. Wang, X. Wang and J. Bai, *J. Cryst.Growth* **311**, 3445 (2009).
19. H. Bi, X. Wang, H. Li, B. Xi, Y. Zhu and Y. Qian, *Solid State Commun.* **149**, 2115 (2009).
20. L. Han, Z. Shan, D. Chen, X. Yu, P. Yang, B. Tu and D. Zhao, *J. Colloid Interf. Sci.* **318**, 315 (2008).
21. S. W. Cao and Y. J. Zhu, *Nanoscale Res. Lett.* **6**, 1 (2011).
22. B. Liu, W. Zhang, Q. Zhang, H. Zhang, J.Yu, X. Yang, *J. Colloid Interf. Sci.* **375**, 70 (2012).
23. L. Leon, A. Bustamante, A. Osorio, G. S. Olarte, L. S. Valladares, C. H. W. Barnes and Y. Majima, *Hyperfine Interact.* **202**, 131 (2011).
24. B. Wang, J. S. Chen, H. B. Wu, Z. Wang, and X. W. Lou, *J. Am. Chem. Soc.* **133**, 17146 (2011).
25. G. Sharma and P. Jeevanandam, *RSC Adv.*, **3**, 189 (2013).
26. S. Zhao, H. Y. Wu, L. Song, O. Tegus and S. Asuha, *J. Mater. Sci.* **44**, 926 (2009).
27. Y. Zhu, F. Y. Jiang, K. Chen, F. Kang and Z. K. Tang, *J. Alloys and Compds.* **509**, 8549 (2011)
28. V. Sreeja and P. A. Joy, *Mater. Res. Bull.*, **42**, 1570 (2007).
29. H. M. Lu and X. K. Meng, *J. Phys. Chem. C*, **114**, 21291 (2010).

Mater. Res. Soc. Symp. Proc. Vol. 1547 © 2013 Materials Research Society
DOI: 10.1557/opl.2013.566

Multifunctional silicone nanocomposites for advanced LED encapsulation

Ying Li, Peng Tao, Richard W. Siegel, and Linda S. Schadler
Department of Materials Science and Engineering and Rensselaer Nanotechnology Center,
Rensselaer Polytechnic Institute, Troy, NY 12180, U. S. A.

ABSTRACT

The addition of high refractive index (RI) inorganic nanoparticles (NPs) to LED
encapsulation materials can lead to higher light extraction efficiency. In addition, the NPs can be
carriers for additional functionality such as color conversion. Using a simple "grafting-to"
approach, bimodal polydimethylsiloxane (PDMS) brushes were grafted onto high-RI ZrO_2 NPs.
Subsequently, an organic phosphor, 6-[fluorescein-5(6)-carboxamido]hexanoic acid (FCHA),
was attached onto the PDMS-grafted ZrO_2 NPs via a facile ligand exchange process. The
bimodal polymer brush design enables homogenous dispersion of the surface functionalized NPs
within the silicone matrix. The functionalized NPs with ~53 wt% ZrO_2 core have a ~0.08 higher
RI than neat silicone, and the NP-filled silicone nanocomposites exhibit a transparency of ~ 90%
in the 550-800 nm wavelength range. In addition, the nanocomposites could be excited at a
wavelength around 455 nm by a blue LED and undergo secondary yellow emission at around
571 nm. It is expected that the prepared nanocomposites can be used as high-efficiency, non-
scattering, color-tuned materials for advanced LED encapsulation.

INTRODUCTION

Compared to epoxy-based LED encapsulants, which tend to yellow over time with
exposure to high operating temperatures and/or absorption of UV-blue light, silicone resins have
higher photochemical and thermal stability, high transparency in the UV-visible region, low
water permeability, and tunable hardness.[1] Silicone encapsulants would open up exciting new
luminaire designs and allow for penetration of phosphor-converted LED lamps into the solid-
state-lighting market if they had a higher RI and better color-conversion properties.[2, 3]
Increasing the RI increases the angle of the light-escape cone, thereby enhancing the light
extraction efficiency.[4] Adding non-scattering color conversion properties to the encapsulant
opens up new luminaire geometries. Currently, the most commonly used phosphors are
composed of an inorganic host substance, such as yttrium aluminum garnet (YAG), doped with
rare-earth elements.[5] With increasing concerns over the resource depletion of rare-earth
elements, organic fluorescent materials have attracted attention because of their low cost, ease of
fabrication, color tuning via modifying π–π^* transitions through molecular/structure design,
solubility in organic solvents enabling molecular-level doping into polymers, and generally good
compatibility with polymer matrices.[3]

In this work, we have demonstrated the preparation of multifunctional silicone
nanocomposites with combined high RI and color conversion functionality by uniformly
dispersing organic-phosphor functionalized high-RI ZrO_2 NPs within a silicone matrix. In order
to achieve higher NP loading, and thus higher RI of the nanocomposite, while minimizing
transparency lose due to Rayleigh scattering as well as fluorescent quenching, homogenous
dispersion of the NPs is critical. Conventional attempts to use monomodal (single population)
grafted polymer brushes to control nanoparticle dispersion are challenged by the need for high

graft density to shield particle core-core attractions and the need for low graft density to facilitate the penetration of matrix chains into the brush to suppress dewetting.[6, 7] In the present study silicone compatible PDMS brushes with two significantly different molecular weights were grafted to the surface of ZrO_2 NP cores to compatibilize the nanoparticle, both enthalpically and entropically. The yellow-emitting FCHA dye was subsequently attached onto the grafted NPs via a facile ligand exchange process to introduce color-conversion functionality. The resulting silicone nanocomposites with tunable interfacial properties and optical functionalities enable new opportunities for advanced LED packaging.

EXPERIMENT

Spherical ZrO_2 NPs with a radius of ~1.9 nm were synthesized using a non-aqueous, surfactant free synthetic approach.[8] In a typical synthesis, 2.22 g of zirconium isopropoxide isopropanol complex (98%, Alfa Aesar) was dissolved in 30 mL of benzyl alcohol (99%, Sigma Aldrich), and the mixture was transferred to a 45 mL stainless steel pressure vessel (Parr), which was heated to 240 °C. After 4 days, a white turbid suspension was retrieved from the cooled vessel and wet ZrO_2 NPs were isolated by centrifugation at 10,000 rpm for 10 min and re-dispersed in chloroform. Silicone compatible PDMS brushes were synthesized through direct modification of commercial hydroxyl-terminated PDMS (Gelest) into a phosphate-terminated PDMS based on a method reported in the literature.[9] The phosphate head group replaces the weakly bonded capping ligands on the as-synthesized NP surfaces. The phosphate-terminated PDMS was added to the transparent NP chloroform solution and refluxed under stirring for 24 h to complete the "grafting-to" process. The grafted NPs were washed using methanol and re-dispersed in chloroform.

FCHA (Sigma-Aldrich) was used to introduce color-conversion capability into the nanocomposite system. FCHA was completely dissolved in mixtures of chloroform and dimethylformamide (DMF), and then added into the grafted NP chloroform solution in a bath sonicator. The reaction mixture was sonicated for 30 min to facilitate the attachment of organic phosphor molecules via the carboxylic acid anchoring group. The phosphor-functionalized NPs were then washed and re-dispersed in chloroform. The functionalized NP chloroform solution was subsequently mixed with a silicone resin (Gelest). The mixture was homogeneously stirred and then put in a vacuum oven overnight for solvent removal at room temperature. After complete solvent removal, the nanocomposite was cured at 120 °C.

In a typical TGA analysis, the NP sample was heated from 30 °C to 800 °C under a N_2 flow at a heating rate of 10 °C/min. The refractive index of the nanocomposites was measured using variable angle spectroscopic ellipsometry (VASE, J.A Woollam Co.) on a spin-coated sample with a thickness in the range of 50 to 100 nm on a Si wafer. The measured results were fitted with the Cauchy model. The nanocomposite samples applied on glass slides (~1 mm thick) were used to measure optical transparency (UV-vis, Perkin–Elmer Lambda 950) and luminescence spectra (F-4500, Hitachi).

RESULTS AND DISCUSSION

The morphology and nanocrystal structure of the as-synthesized ZrO_2 NPs are shown in Figure 1. The TEM characterization shows homogeneously distributed, near monodisperse ZrO_2 NPs with an average radius of ~1.9 nm. All the peaks in the XRD pattern can be assigned to the

ZrO₂ cubic phase (JCPDS, 27-997). The small-size of the ZrO₂ NPs is desirable for reducing transparency loss due to Rayleigh scattering. Also, the relatively large band gap of ZrO₂ (~5 to 7 eV) compared to other metal oxide nanofillers, such as TiO₂ (~3.5 eV), leads to less photocatalytic activity and thus better polymer stability.[10, 11]

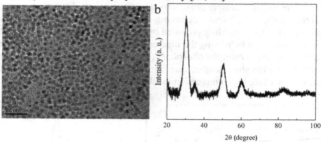

Figure 1. (a) TEM image of as-synthesized ZrO₂ NPs, and (b) their XRD pattern.

The bimodal PDMS brush grafted NPs were then prepared using a two-step "grafting-to" approach. The phosphate-terminated PDMS brushes robustly anchor onto ZrO₂ NPs through the strong binding of organo-phosphate with metal oxides. After attaching the long brush in the first step, the short brush filled in the remaining space on the particle surface at a higher graft density. As illustrated in Figure 2a, the FCHA phosphor molecule was then attached to the grafted NPs via the coupling of the carboxylic acid head group with the NP surfaces. As shown in Figure 2b, the transparent yellow-orange chloroform solution of the functionalized NPs is preliminary evidence of the successful attachment of FCHA, since the pure FCHA phosphor cannot be dissolved in chloroform as free molecules. The change in weight percentage of the functionalized NPs shown in the normalized TGA curves further confirms the existence of the ligand exchange process during functionalization, and the increased weight loss during the TGA analysis indicates the successful attachment of FCHA.

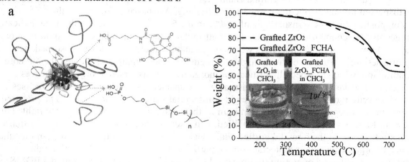

Figure 2. (a) Schematic illustration of a functionalized NP with bimodal PDMS polymer brushes and FCHA molecules attached. Gray polymer chains represent long brushes (M_w=36 kg/mol) and lighter gray chains represent short brushes (M_w=10 kg/mol). (b) Normalized TGA curves for PDMS grafted ZrO₂ NPs before and after FCHA functionalization.

Figure 3 shows an increase of RI (~0.08) over neat silicone for silicone containing functionalized NPs with ~53 wt% ZrO_2 core. The solution absorption and photoluminescence (PL) emission spectra of the FCHA-functionalized NPs are nearly the same as those of the free FCHA, except for the strong absorption of UV light attributable to the ZrO_2 core. Since the optical properties of the organic phosphors are very sensitive to their dispersion state, the unaltered PL spectrum indicates a homogeneous distribution of the organic phosphor molecules on the NP surface and effective shielding provided by the matrix-compatible PDMS brushes. White light is usually produced by mixing the blue emission (~460 nm) from InGaN and the yellow luminescence converted from the phosphors (~560 nm). Therefore, the PL behavior of the functionalized NP solution suggests that such high-RI, FCHA-functionalized NPs would be promising candidates for white light conversion in LEDs.

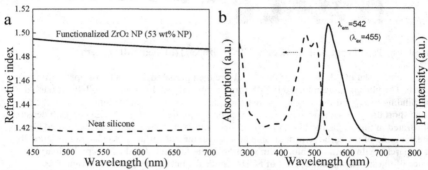

Figure 3. (a) Comparison of RI dispersion of neat silicone and silicone with functionalized ZrO_2 NPs. (b) Solution PL emission and UV–Vis absorption spectra of functionalized NPs.

With the bimodal PDMS grafted chain design, the organic phosphor functionalized NPs were dispersed into a silicone matrix at 30 wt% loading. As shown in Figure 4a, the nanocomposite exhibits a transparency of ~ 90% in the 550-800 nm wavelength range, indicating homogenous dispersion of functionalized NPs attributable to the bimodal surface modification. Below 550 nm, the organic phosphors were excited and the incident light was absorbed. The short brushes are grafted at relatively high graft densities and enthalpically screen the NP core-core attraction, which is especially critical for inorganic NPs dispersed in organic matrices considering their large surface energy mismatch. The sparsely grafted long brush suppresses entropic dewetting of high-molecular-weight commercial silicone matrices.[6, 7] The prepared transparent silicone nanocomposite can minimize the scattering loss of the converted light. This is superior to the state-of-the-art silicone filled with micron-sized inorganic phosphor particles. Another advantage of the bimodal grafted chain design is the easy tuning of the inter-molecular distance of organic phosphor molecules on the particles. It can be expected that the light emission luminosity and operating lifetime can be tailored by optimizing the bimodal PDMS brush design. The maximum PL intensity of the nanocomposite was observed at 571 nm, which is yellow in color. The red-shifted PL spectrum of the nanocomposite (~ 30 nm), compared to the functionalized NPs in chloroform (Figure 3b), is probably attributable to the stronger interactions between fluorescein units in the solid state.[2] However, the strong absorption at 400-550 nm wavelength suggests that the phosphor concentration in the current nanocomposites might be too high for blue LED encapsulation, since they completely absorbed the blue emission from the LED chip. For future white LED applications, either the loading of functionalized NPs or the number of phosphor molecules on each NP should be reduced, which could be achieved by increasing the graft density of the PDMS brushes.

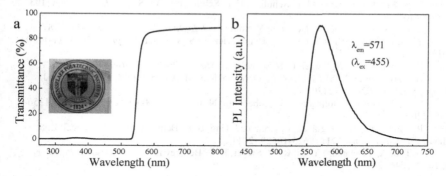

Figure 4. (a) UV-vis transmittance spectrum (actual sample shown in inset) and (b) PL emission spectrum of functionalized silicone nanocomposite with 30 wt% loading of organic phosphor functionalized ZrO_2 NPs.

CONCLUSIONS

High RI ZrO_2 NPs were synthesized, surfaced modified with bimodal PDMS polymer brushes, and then functionalized with organic phosphor molecules. The functionalized NPs achieved homogenous dispersion within the silicone matrix. The high-RI and color-converting functionalities are introduced in the silicone nanocomposite without sacrificing the good transparency and flexibility of the silicone resin. The bimodal surface ligand design provides an effective tool for compatibilizing the inorganic nanofiller with the organic matrix. The multifunctional silicone nanocomposite exhibits desired fluorescent properties for color-converting in LED encapsulation. By varying the graft density of the bimodal PDMS brushes and functionalized NP loading, the PL property of the multifunctional silicone nanocomposite can be further optimized for white LED applications.

ACKNOWLEDGMENTS

This work was supported by the Engineering Research Centers Program of the National Science Foundation under Cooperative Agreement EEC-0812056 and by NYSTAR under contracts C080145 and C090145 and by the Nanoscale Science and Engineering Initiative of the National Science Foundation under NSF award number DMR-0642573.

REFERENCES

1. L. D. Boardman; D. S. Thompson; C. A. Leatherdale U.S. Patent 0199291 A1, (2006).
2. P. Uthirakumar; E. K. Suh; C. H. Hong; Y. S. Lee, *Polymer* **46**, 4640-4646 (2005).
3. L. Zhang; B. Li; B. Lei; Z. Hong; W. Li, *J. Lumin.* **128**, 67-73 (2008).
4. F. W. Mont; J. K. Kim; M. F. Schubert; E. F. Schubert; R. W. Siegel, *J. Appl. Phys.* **103**, 1-6 (2008).
5. K. Bando; K. Sakano; Y. Noguchi; Y. Shimizu, *J. Light and Vis. Env.* **22**, 2-5 (1998).
6. Y. Li; P. Tao; A. Viswanath; B. C. Benicewicz; L. S. Schadler, *Langmuir* **29**, 1211-1220 (2013).
7. P. Tao; Y. Li; R. W. Siegel; L. S. Schadler, *J. Mater. Chem. C* **1**, 86-94 (2013).
8. G. Garnweitner; L. M. Goldenberg; O. V. Sakhno; M. Antonietti; M. Niederberger; J. Stumpe, *Small* **3**, 1626-1632 (2007).
9. M. A. White; J. A. Johnson; J. T. Koberstein; N. J. Turro, *J. Am. Chem. Soc.* **128**, 11356-11357 (2006).
10. P. Tao; Y. Li; A. Rungta; A. Viswanath; J. Gao; B. C. Benicewicz; R. W. Siegel; L. S. Schadler, *J. Mater. Chem.* **21**, 18623-18629 (2011).
11. P. Tao; A. Viswanath; Y. Li; R. W. Siegel; B. C. Benicewicz; L. S. Schadler, *Polymer* **54**, 1639–1646 (2013).

Mater. Res. Soc. Symp. Proc. Vol. 1547 © 2013 Materials Research Society
DOI: 10.1557/opl.2013.565

Study on the Production of Silver/Modified Clay Nanocomposites

Natália F. N. Pessanha[1] and Gerson L. V. Coelho[1]

[1] Separation Process Laboratory, Department of Chemical Engineering, Federal Rural University of Rio de Janeiro, Seropédica, Rio de Janeiro, Brazil.

ABSTRACT

The aim of this study was to investigate the application of modified clay as a support in the synthesis of silver nanoparticles. Silver nitrate ($AgNO_3$) was used as the silver precursor in several concentrations (0.005 M, 0.01 M, 0.02 M, 0.05 M, and 0.1 M) to obtain Ag-MMT purified and modified clay nanocomposites. The properties of nanocomposites were also studied as a function of the concentration of the reducing agent, sodium borohydride ($NaBH_4$). It was observed through X-ray Diffraction that the MMT purified structure was gradually exfoliated with increased concentrations of $AgNO_3$, while the modified clay structure remained intact. As observed through UV-vis spectra, samples of Ag^+-MMT were reduced with $NaBH_4$ to produce Ag^0 and its particle diameter is dependent on the concentration of $NaBH_4$.

INTRODUCTION

Silver (Ag) has been recognized for its oligodynamic activity since ancient times and its effects on cellular activity in living organisms even at low concentrations. The use of Ag as an antimicrobial material proved to be advantageous because it presents the highest degree of toxicity to more than 650 pathogenic microorganisms and still shows low toxicity to animal cells. It has been observed throughout the years that the antimicrobial activity of Ag was enhanced when it is presented in the colloidal form, i.e., in nanometer sizes (from 10^{-9} m to 10^{-6} m). Nano Ag provides increased number of particles per unit area [1,2,3]. The preparation of Ag nanoparticles consists mostly of two steps: reduction of Ag cations mainly from $AgNO_3$ solutions and stabilization of the resulting Ag nanoparticles. Therefore, clay montmorillonite (MMT) has a lamellar structure, swelling capacity in aqueous media, and cation exchange capacity. The lamellar surface has charge-compensating cations (Ca^{2+}, Na^+ and Li^+), which can be exchanged for other metal ions. The reduction of these metal ions leads to nano-sized particles because the lamellar spacing serves as a nano-reactor limiting the size of particles preventing agglomeration, and ensuring optimal dispersion of particles on the clay's surface [4]. Given that Ag-MMT nanocomposites have been produced, and considering its known antimicrobial activity, the aim of this study was to synthesize Ag nanoparticles in modified clay to increase the basal spacing with a quaternary salt that develops higher adsorption capacity of nanometer metallic silver.

EXPERIMENTAL DETAILS

The commercial sodium bentonite clay (Argel CN 35) from Brazil has a cation exchange capacity (CEC) of 100 meq/100 g, was used in this study, and codified as MMT. The synthesis of nanocomposite comprises five steps: clay purification and modification, Ag adsorption and reduction, and characterization of the materials produced during these steps. Firstly, bentonite clay was purified with hydrogen peroxide for 24 hours to remove organic contaminants. Subsequently, the excess peroxide was removed through a thermostatic heating at 80 °C, and the purified clay decanted and dried at 60 °C. Clay modification was performed using a solution of cetyltrimethylammonium bromide under constant stirring for 24 hours in order to increase the basal spacing of MMT [5]. The adsorption process involved the addition of 200 mg of clay in 200 mL of 0.005 M, 0.01 M, 0.02 M, 0.05 M, and 0.1 M $AgNO_3$ followed by vigorous agitation for 24 hours. The suspension was centrifuged afterwards for 20 minutes, filtered, and dried at 50 °C for 24 hours. The clay, impregnated with Ag ions (Ag^+-MMT), was mixed in a solution of $NaBH_4$ 0.001 M, 0.005 M, and 0.01 M that was freshly prepared to promote Ag^+ ions reduction; this mixture was subjected to agitation for 1 hour, centrifuged, washed with deionized water, and dried at 50 °C [6].

The characterization of modified and unmodified clay was performed using infrared spectroscopy (FT-IR Perkin Elmer Spectrum 100) to investigate structural changes that occur because of changes in the clay mineral. The purified and modified clay and nanocomposites were analyzed with an X-ray diffractometer (Miniflex Rigatu) using $CuK\alpha$ radiation, 40 kV, 30 mA, in the angle range of $2° < 2\theta < 55°$. The basal spacing (d_{001}) was calculated through the Bragg's Law (Equation 1) and the size of the Ag nanoparticles, present in the structure of clay, was calculated using the Scherrer equation (Equation 2) where θ is the angle of diffraction, λ is the wavelength (0.15418 nm), n is the order of diffraction equal to unity, k is a constant, and β is a parameter that corresponds to the measurement of the width of a diffraction peak of 2θ where the intensity falls to half (FWHM - full width at half maximum). The colloid of silver nanoparticles, obtained after reduction of Ag^+, was analyzed with a UV-visible spectrophotometer (BEL Photonics 1105) using wavelengths in the range of 320-600 nm.

$$\lambda\, n = 2\, d_{001}\, sen\theta \qquad (1)$$

$$D\, k\, \lambda = \beta\, cos\theta \qquad (2)$$

DISCUSSION

Organic matter has great influence on the exchange capacity of clays; it can act as a protective colloid hampering the exchange of cations from clay minerals [4]. The infrared spectra of the purified and modified clay are shown in Figures 1 and 2, respectively. Both samples of montmorillonite clay presented bands in the region of approximately 3631 cm^{-1} and 3449 cm^{-1}, which are attributed to the stretching vibration of the OH group and structural adsorbed water, respectively. The bands on the structure of the clay, in both samples, are observed in the 1043 cm^{-1} region attributable to the stretching vibration of the Si-O tetrahedral layer, while the 845 cm^{-1} and 523 cm^{-1} bands are related to angular Si-O-al (octahedral layers of aluminosilicate) vibrations and 464 cm^{-1} angular Si-O-Si vibrations [7,8].

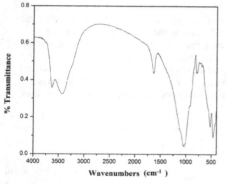

Figure 1: Infrared spectrum of the purified MMT.

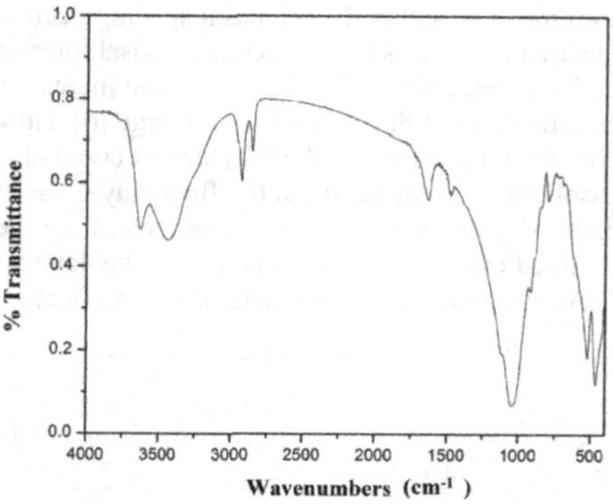

Figure 2: Infrared spectrum of the modified MMT.

Figure 2 shows the absence of bands associated with the stretching of the CH_2 and CH_3 groups related to organic matter at 2923 cm^{-1} and 2852 cm^{-1}. The presence of quaternary salt in the modified clay (Figure 2) was indicated by the appearance of two intense bands at 2930 cm^{-1} and 2847 cm^{-1}, which correspond to the asymmetric and symmetric vibrational modes of the CH_2 group, respectively. In addition, the formation of a band at 1475 cm^{-1} relative to the asymmetric deformation of the CH was observed [8,9].

Table 1 present values of the basal spacings for each sample calculated using the Bragg's Law after the X-ray diffractograms analysis. We observed increased spacing after the process of clay purification and modification, thus, confirming the efficiency of the adopted procedures. The change corresponded to the replacement of exchangeable ions present in the galleries of the clay by the HDTMA$^+$ organic cation, cetremide, with the aims to expand the clay's basal spacing. Therefore, a satisfactory differentiation in the basal spacing of modified clay (1.41 nm) was obtained when compared with commercial clay (1.27 nm).

Table 1: Basal spacings.

Sample	2θ	d_{001} (nm)
Commercial MMT	6,96	1,27
Purified MMT	6,44	1,36
Organophilic MMT	6,26	1,41

The X-ray diffractograms containing basal spacings of the purified and organophilic clay that were used for a structural analysis of materials after adsorption of silver cations are shown in Figures 3 and 4, respectively. The diffractogram for the purified clay (Figure 3) showed reduction in the intensity of the peak (001) with increasing concentration of silver nitrate solution. Consequently, the basal spacing decreased, which was originally 1.36 nm after the purification process. The decreased basal intensity of diffraction indicated that the lamellar clay had been partially exfoliated due to the increase in Ag ions in the clay's galleries causing a structural stress during cation exchange [6]. However, the intensity in the diffractograms at the concentration of 0.1 M AgNO₃ almost completely disappears indicating an exfoliation/delamination in the final clay. In contrast the structure of the modified clay was not altered by the increased concentrations of Ag^+ because the presence of the $HDTMA^+$ ion is retained in the interlayer causing the clay layers to remain joined and arranged in parallel, thus, preventing the exfoliation/delamination effect.

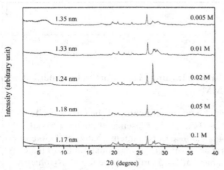

Figure 3: X-ray diffractogram of purified clay intercalated with silver ions in different concentrations of AgNO₃ solutions.

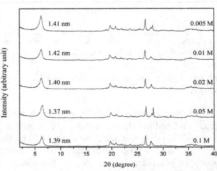

Figure 4: X-ray diffractogram of modified clay with silver ions in different concentrations of AgNO₃ solutions.

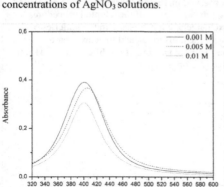

Figure 5: UV-visible spectrophotometry of supernatants after the process of silver ions reduction present in purified clay in different NaBH₄ concentrations.

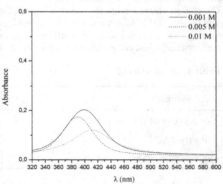

Figure 6: UV-visible spectrophotometry of supernatants after the process of silver ions reduction present in modified clay in different NaBH₄ concentrations.

The materials produced using the concentration of 0.01 M AgNO$_3$ were taken as the basis for the reduction of Ag ions adsorbed in the structure of purified and modified clay. Caramel coloration was observed at all concentrations, differing only in transparency, after the reduction process with three concentrations of NaBH$_4$ in the solutions containing purified clay. However, the reduction of the ions present in the samples of modified clay, using solutions containing NaBH$_4$ at 0.001 M, 0.005 M, and 0.01 M, showed products of pale yellow and light yellow coloration, and colorless, respectively. This difference between purified and modified clay is related to a higher adsorption capacity of Ag ions in clay that had been purified by replacing the exchangeable ions present in the gallery, which is absent in clay modified by Ag$^+$. The reduction caused an increase in the concentration of nanoparticles in solution, showing a small agglomerative state, which was independent of the used NaBH$_4$ concentration.

Analysis by UV-visible spectrophotometry (Figures 5 and 6) showed peaks between 380 and 420 nm, which characterized a mean particle size in the range of 5-50 nm in solution [10]. A reduction in the intensity of the peaks with increasing concentration of NaBH$_4$ was also observed. This result indicates that particles with smaller diameters are synthesized at higher NaBH$_4$ concentrations, which favors the adsorption of nanoparticles on the clay's surface. However, in Figure 6, the spectra of the 0.001 M and 0.01 M concentrations showed half reduced the amount of nanosized particles in solution. This reduction indicates that a possible greater adsorption of modified clay nanoparticles occurred because there was no colloid agglomeration and the particles formed presented smaller diameter. Ag remains synthesized on solid phase (MMT) during the process of its reduction to metallic Ag, while Ag ions are removed from the interlayer and then reduced and adsorbed on the outer surface and edges of the clay.

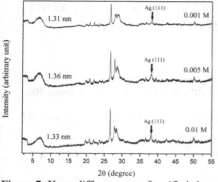 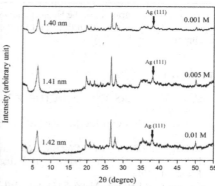

Figure 7: X-ray diffractogram of purified clay after reduction of silver ions adsorbed in NaBH$_4$ solutions.

Figure 8: X-ray diffractogram of modified clay after reduction of silver ions adsorbed in NaBH$_4$ solutions.

The X-ray diffractograms of purified Ag-MMT and modified Ag-MMT nanocomposites, presented in Figures 6 and 7, respectively, showed that the formation and adsorption of Ag nanoparticles did not modify the structure of the modified clay, therefore, their basal spacings remained similar since the adsorption of cations in the reduction process. However, we observed an increase in basal spacing in the clay purified by reduction with NaBH$_4$ at 0.005 M and 0.01

M. This expansion results from the intercalation of Na (from the NaBH$_4$ solutions), which provides the restoring stacked lamellar structure.

All nanocomposites showed diffraction peaks at approximately 38° in 2θ related to Ag crystals in the face-centered cubic crystallographic planes (111). The average size of crystallites present in the Ag nanocomposites obtained by reduction of Ag$^+$ with 0.01 M NaBH$_4$ was calculated because these were the samples that maintained their basal spacings constant, after the reduction process. The average Ag crystallite size, namely D$_{111}$, obtained for the Ag-MMT purified sample through the Scherrer equation, was 7.59 nm in the diffraction angle of 38.12°. The D$_{111}$ of the Ag-MMT modified sample was 7.67 nm in the diffraction angle of 38.02°.

CONCLUSIONS

The modified clay showed a positive result in production of silver/modified clay nanocomposites because its structure remained intact after the adsorption and reduction of silver ions. Moreover, the modified clay showed higher adsorption capacity of silver nanoparticles present in the colloid because these nanoparticles are smaller than the default of agglomerates when compared to the result obtained from the purified clay. The results demonstrated that the combination of Ag-MMT nanocomposite with modified clay with various kinds of material is advantageous and valid for future industrial applications.

ACKNOWLEDGMENTS

The authors are grateful to the Heterogeneous Catalysis Laboratory DEQ/IT/UFRRJ and the Center for Catalysis from COPPE/UFRJ for the assistance with the analyses, which were of great importance for this study.

REFERENCES

1. C. W.CHAMBERS, C. M. PROCTOR, P. W. KABLER, J Am Water Works Assoc, 208-216 (1962).
2. J. P. GUGGENBICHLER, M. BOSWALD, S. LUGAUER, T. KRALL, J Infect, S16-S23 (1999).
3. R. DASTJERDI, M. MONTAZER, Coll. Surf. B, 79, 5-18 (2010).
4. P. S. SANTOS, *Ciência e Tecnologia de Argilas*, 2nd ed., vol.1 (Edgard Blucher Ltda, S. Paulo, SP, 1989).
5. G. L. V.COELHO, F. AUGUSTO, J. PAWLISZYN, Ind. Eng. Chem. Res., 40, 364-368 (2001).
6. P. PRAUS, M. TURICOVÁ, M. VALÁŠKOVÁ, J. Braz. Chem. Soc, 19, 549-556 (2008).
7. J. MADEJOVÁ, M. JANEK, P. KOMADEL, H. J. HERBERT, H. C. MOOG, Appl Clay Sci, 20, 255-271 (2002).
8. L. B. PAIVA, A. R. MORALES, F. R. V. DÍAZ, Appl Clay Sci, 42, 8-24 (2008).
9. H. HONGPING, F. L. RAY, Z. JIANXI, Spectrochim. Acta, Part B, 60, 2853-2859 (2004).
10. S. SOLOMON, M. BAHADORY, A. JEYARAJASINGAN, S. UTKOWSKY, C. BORITZ, J. Chem. Educ, 84, 322-325 (2007).

Mater. Res. Soc. Symp. Proc. Vol. 1547 © 2013 Materials Research Society
DOI: 10.1557/opl.2013.637

DFT Investigation of the Mechanism and
Chemical Kinetics for the Gelation of Colloidal Silica

Steven S. Burnett[1] and James W. Mitchell[1]
[1] Howard University 2300 Sixth Street, Rm. 1124, NW, Washington, DC 20059 U.S.A.

ABSTRACT

The mechanism for the gelation reaction of colloidal silica, $Si(OH)_4$ $+Si(OH)_3 (O)^- \dashrightarrow Si_2O_8H_5^- + H_2O$, by an anionic pathway was investigated using density functional theory(DFT). Using transition state theory, the rate constants were obtained by analyzing the potential energy surface at the reactants, saddle point, and the products. In addition, reaction rate constants were investigated in the presence of ammonium chloride (NH_4Cl) and sodium chloride (NaCl). These salts act as catalysts to induce gelation by destabilizing the double layer of colloidal silica to allow for Van der Waal interactions. Furthermore, it was observed that ammonium chloride plays an important role by initiating a hydride transfer allowing the reaction to proceed from the second transition state to the final product.

INTRODUCTION

Chemical processing to generate nanocomposites is a forefront approach to producing novel materials. In particular, the gelation technique, utilized by the sol-gel process, provides an effective means to achieve highly dispersed fluid phase composites [1–3]. For example, the chemistry behind silica based gelation techniques has been studied extensively experimentally and theoretically [4–7]. To illustrate, one of the earliest theoretical approaches for studying the kinetics of sol-gel processing characterized the condensation reaction from a donor-acceptor approach. Additionally, spectroscopic techniques (FTIR, ^1H NMR, and ^{29}Si NMR) have been utilized extensively for determining the influence of kinetics on the micro-structure of the gel [8,9].

Silica based gelation frequently proceeds by one of two pathways, the neutral and the anionic pathway. However, the anionic pathway is more favorable and has a lower activation barrier [7][10][11]. To illustrate, the neutral pathway proceeds from reactants to products through a single transition state structure, whereas, the anionic pathway requires two transition state structures [12]. Identically, both pathways are connected on the reaction path by a five-coordinate SiO-Si intermediate.

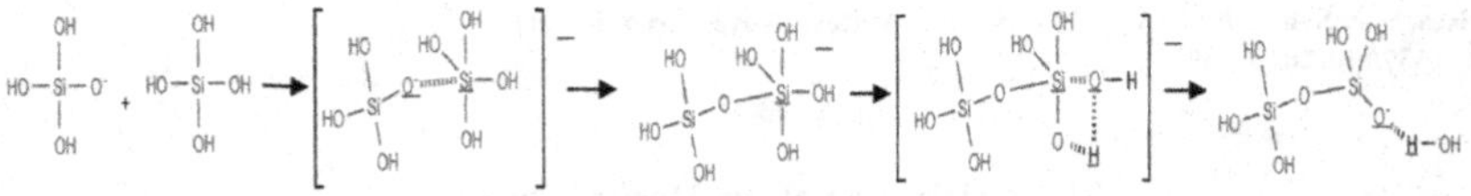

Scheme 1: The anionic mechanism for silica gelation

Furthermore, the pH of the medium determines which pathway is available. In a high pH environment, the anionic pathway will be dominant and is preferred as the deprotonated SiO− is more stable than the protonated $SiOH_2+$ group [11][13–15]. Consequently, the deprotonated SiO− species serves as an initiator for the gelation reaction. In addition to pH being a key parameter for silica gelation, understanding and controlling the zeta potential is paramount for colloidal chemistry. Namely, salts have the capability to alter the zeta potential by reducing the electric double layer surrounding the dispersant(s), and therefore catalyzing the gelation by facilitating Van der Waal attraction leading to gelation.

Computational Experiment

Ab initio quantum chemical calculations were carried out using the density functional theory (DFT) method with the B3LYP hybrid exchange-correlation functional [16][17]. The B3LYP functional was chosen because of its ability to provide reliable values for vibrational frequencies and activation energies for various reaction profiles [7][18][19] in addition, this approach is practical because of its ability to provide accurate energy calculations with tolerable cpu cost [19]. The 6-31G+(d,p) basis set was used to expand the molecular orbital to optimize the equilibrium geometries, and for preforming saddle point searches in the gas phase. As a result, this smaller basis set reduced the cpu time needed to calculate the required Hessian matrix to preform saddle points searches. The requirement for finding the saddle point is that one of the eigenvalues of the Hessian matrix is negative.

Next, single-point energy calculations were carried out with a larger basis set, 6-311++G(d,p), to provide a more complete correlation treatment on the optimized geometries and saddle points. In addition, we included the effect of the solvent on our calculations with the 6-311++G(d,p) basis set by using the COSMO (conductor-like screening model) method implemented in the Gamess [20] *ab initio* package. This approach follows the work reported by Trinh et al [7] about how the bond distances for the equilibrium geometries in the gas phase were always less than 0.01 Å to their COSMO optimizations, therefore, reducing cpu time for our calculations. Lastly, the rate constants were determined from Eyring's transition state theory (TST) from equations 1 and 2. Additionally, tunneling corrections to the TST were calculated using equation 3:

$$k_{rate} = w \frac{K_B T}{h} e^{\frac{-\Delta G}{RT}} \qquad (1)$$

$$\Delta G = \Delta G_{TS} - \Delta G_{reactant} \qquad (2)$$

$$w = 1 + \frac{1}{24} \frac{hv}{K_B T} \qquad (3)$$

where w is the tunneling correction term, K_B is the Boltzmann constant, h is the Planck constant, R is the gas constant, T is the temperature, and v is the imaginary frequency at the transition state.

RESULTS and DISCUSSION

Reaction Mechanisms

Results for the gelation reaction of silica and silica with salts are displayed in Table 1. It is important to note that there is a change in energy when going from the gas phase to solvent by COSMO; the change for the silica reaction complex (RC1) was -61.308 Kcal/mol, indicating the importance of solvation effects. Relevant structural data obtained for the first and second transition states are shown in Table 2. The reaction coordinate of the anionic approach to the condensation reaction of two silica molecules is outlined in Figure 1, where the reaction proceeds through two transition states (Figure 2) connected by a hypervalent intermediate (RC2). In the first step of the reaction, the $Si(OH)_3-$ species is stabilized by hydrogen bonding provided by $Si(OH)_4$, and the resulting transition state corresponds to the hypervalent SiO−Si bond being formed. Additionally, the rate determining step is the dissociation of OH in the second transition state where the OH− ion deprotonates the monomeric species to yield $Si(OH)_3O-$.

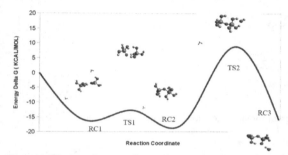

Figure 1: The reaction coordinate diagram for silica gelation

(a) TS 1   (b) TS 2

Figure 2: The transition state geometries for the gelation reaction

Next, the influence of two representative counter ions in the Hofmeister sequence, NaCl and NH₄Cl, (Figures 5 and 6), were used to model the gelation reaction and relate their effect on destabilizing the electric double layer for silica sols. The change in the total free energy for TS1 and TS2 for both NaCl and NH₄CL demonstrate their effectiveness as catalyst by lowering the energy when compared to pure silica (Table 1). Furthermore, it is apparent that more energy is needed to reach the second transition state with NaCl (Figure 3), but requires less energy for the first transition state. This provides some insight into colloidal chemistry where salts are used to destabilize the electric double layer, therefore increasing the Van der Waals forces to allow for successful collisions from silica molecules to aggregate and form a gelled network. However, reaction kinetics dictates that the rate limiting step determine the overall kinetics. Therefore, when we look at NH₄Cl (Figure 4), its second transition state is lower than that of NaCl, and its final product (RC3) has a more negative change in energy.

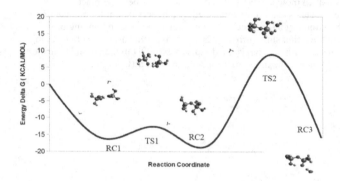

Figure 3: The reaction coordinate diagram for silica gelation in the presence of NaCl

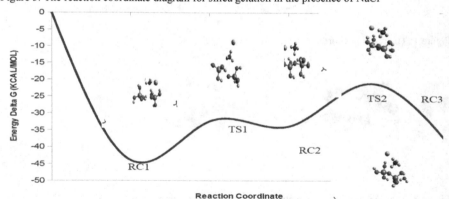

Figure 4: The reaction coordinate diagram for silica gelation in the presence of NH₄Cl

	RC1	TS1	RC2	TS2	RC3	$TS1rate$	$TS2rate$	$w1$	$w2$
Dimer	-16.28	-12.95	-17.12	8.57	-17.12	1.94E+22	3.24E+6	1	1
NaCl	-39.66	-36.98	-36.77	-18.34	-36.45	8.00E+39	1.73E+26	1	1
NH$_4$Cl	-44	-32.42	-33.94	-21.29	-38.38	3.63E+36	2.52E+28	1	1

Table 1: Calculated $\Delta Energy$ for the total free energy in the solvent (Kcal/mol) for the initial stage of the gelation reactions forming a silicate dimer, dimer with NaCl, and dimer with NH$_4$Cl using the COSMO model for the reactant coordinates (RC) and transition states (TS). Transition state theory was used to calculate the rates for the two transition states ($TS1_{rate}$ and $TS2_{rate}$), and the tunneling correction factors ($w1$ and $w2$).

	$Si_6 - O_9$	$Si_6 - Si_1$	$Si_6 - O_2$	$Si_1 - O_2$	$\Theta SiOSi°$
Dimer TS1	1.669°A	3.855°A	3.091°A	1.582°A	106.7
Dimer TS2	2.500°A	2.944°A	1.693°A	1.644°A	123.8
NH$_4$Cl TS1	1.660°A	4.286°A	2.818°A	1.589°A	152.0
NH$_4$Cl TS2	2.402°A	3.114°A	1.676°A	1.646°A	139.2
NaCl TS1	1.703°A	3.682°A	2.380°A	1.603°A	134.2
NaCl TS2	2.364°A	3.067°A	1.684°A	1.649°A	133.8

Table 2: Selected bond properties of the transition states for the silica dimer, dimer with NaCl, and dimer with NH$_4$Cl

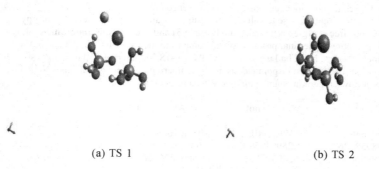

(a) TS 1   (b) TS 2

Figure 5: The transition state geometries for the gelation reaction in the presence of NaCl

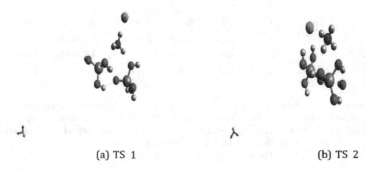

(a) TS 1   (b) TS 2

Figure 6: The transition state geometries for the gelation reaction in the presence of NH$_4$Cl

Reaction Rate Constants

From Equation 1, reaction rate constants in the solvent phase were calculated using the 6-311++G(d,p) basis set for each transition state (Table 1). For all of the gelation reactions, the rate constant for the second transition state is much smaller compared to the first. As a result, from a colloidal chemistry standpoint, the first transition state sheds light on the role of the close proximity needed by the two species to initiate gelation. However, reaching this proximity requires a detailed knowledge of the surface chemistry, in particular the ζ-potential, and its impact on the modification of the surface chemistry.

Thermodynamic Data

Eyring plots (ln k/T vs T^{-1} 10^3) were constructed from the rate constants for the second transition states over the 20◦C to 35◦C range. Figures 7, 8, and 9 relate the influence of the reaction temperature on the gelation of silica, and silica in the presence of the two salts. From a regression analysis, the slope was use to calculate the enthalpy, and the Y-intercept for entropy. As a result, the Gibbs free energies were calculated (Table 3) and used to determine which salt is better for creating a more stable final product. When comparing the two salts, NH$_4$Cl had a ΔG of -21.29 kcal/mol whereas, NaCl had a less negative ΔG of -18.34 kcal/mol. Table 3 suggest that the pure gelation reaction is nonspontaneous at any temperature, whereas NaCl and NH$_4$Cl are spontaneous at low temperatures and all temperatures respectively.

	ΔG kcal/mol	ΔH kcal/mol	ΔS cal/mol
Dimer	8.5700016909	8.5699519018	-0.0001669933
NH$_4$Cl	-21.2898932726	-21.2899455905	0.0001754749
NaCl	-18.3399996055	-18.3401077215	-0.0003626228

Table 3: Gibbs free energies, enthalpy, and entropy extrapolated from the Eyring plots

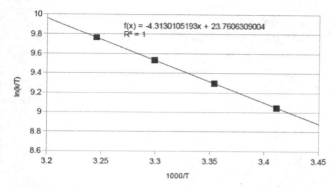

Figure 7: Eyring plot from the rate constants from TS2 of silica over the 20∘C to 35∘C range.

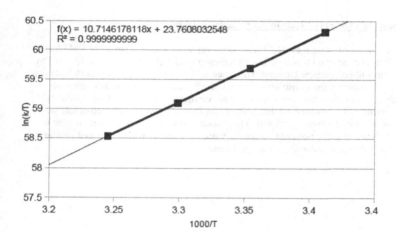

Figure 8: Eyring plot from the rate constants from TS2 of silica with NH_4Cl over the 20∘C to 35∘C range.

179

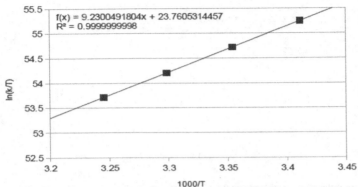

Figure 9: Eyring plot from the rate constants from TS2 of silica with NaCl over the 20°C to 35°C range.

Ammonium Chloride's Observed Role in the Gelation Reaction

Finding the second transition state for NH_4Cl was equivocal when the Hessian at the PM3 level was used. Therefore, instead of a Hessian using a semi-empirical method, one was calculated by a more accurate, but still economical basis set. Huzinaga's 21- split valence basis set (MIDI) was employed, and upon obtaining the transition state, an intrinsic reaction coordinate (minimum energy path) in the forward direction was performed for verification. Utilizing MacMolPlt's [21] visualization capabilities, it was observed that ammonium chloride plays an important role by initiating a hydride transfer allowing the reaction to proceed from the second transition state to the final product (Figure 10). We noticed that for the work by Trinh et al [10] reported similar interactions with *ab initio* molecular dynamic simulations.

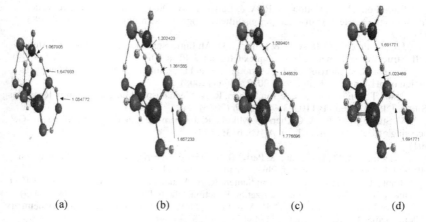

| (a) | (b) | (c) | (d) |

Figure 10: Snap shot of 4 points on the intrinsic reaction coordinate for the gelation reaction indicating the role of NH_4Cl in the final step of the reaction. (a) Represents TS2. Points (b) and (c) are two points on the forward path, whereas point (d) is the last point calculated on the intrinsic reaction coordinate.

CONCLUSIONS

The gelation reaction mechanism and chemical kinetics were investigated by DFT calculations, using the COSMO method, to interpret experimental data for the sol-gel processing of nanocomposites. Furthermore, the effects of specific salts on the gelation kinetics were analyzed to determine their influence on gelation time for processing nanocomposites. The main conclusions from this study are the following: Following the minimum energy pathway from the second transition state, in the presence of NH_4Cl, the disassociation of a water molecule is initiated by a hydride transfer from NH_4Cl. Furthermore, a gelation reaction, in the presence of NH_4Cl is spontaneous at all temperatures, and has a more negative ΔG than NaCl. Lastly, DFT calculations and transition state theory provide detailed knowledge about the optimization of gelation reactions for sol-gel processing.

REFERENCES

[1]   Bhandarkar, S. Sol-Gel Processing for Optical Communication Technology. J. Am. Ceram. Soc 87, 1180–1199 (2004).
[2]   Hench, L. L. & West, J. K. The Sol-Gel Process. Chem. Rev. 1990, 33 – 72 (1961).
[3]   Johann Cho, M. S. P. S., Aldo R. Boccaccini. Ceramic Matrix Composites Containing Carbon Nanotubes. J Mater Sci 44, 19341951 (2009).

[4]   Mora-Fonz, M. J., Catlow, C. R. A. & Lewis, D. W. Oligomerization and Cyclization Processes in the Nucleation of Microporous Silicas. Angew. Chem. Int. Ed. 44, 3082 – 3086 (2005).

[5]   J. C. G. Pereira, C. R. A. C. & Price, G. D. Ab Initio Studies of Silica Based Clusters. Part II. Structures and Energies of Complex Clusters. J. Phys. Chem. A 103, 3268–3284 (1999).

[6]   Nangia, S. & Garrison, B. J. Reaction Rates and Dissolution Mechanisms of Quartz as a Function of pH. Environ. Pollut. 112, 2027 – 2033 (2008).

[7]   Thuat T. Trinh, A. P. J. J. & van Santen, R. A. Mechanism of Oligomerization Reactions of Silica. J. Phys. Chem. B 110, 23099–23106 (2006).

[8]   V. Stanic, T. H. E., A. C. Pierre & Mikula, R. J. Chemical Kinetics Study of the Sol-Gel Processing of GeS2. J. Phys. Chem. A 105, 6136–6143 (2001).

[9]   F. Brunet, M. D., B. Cabane & Perly, B. Sol-Gel Polymerization Studied Through ^{29}Si NMR with Polarization Transfer. J. Phys. Chem. 95, 944–951 (1991).

[10]   Trinh, T. T., Jansen, A. P. J., van Santen, R. A., VandeVondele, J. & Meijer, E. J. Effect of Counter Ions on the Silica Oligomerization Reaction. Chem. Phys. 10, 1775 – 1782 (2009).

[11]   Mora-Fonz, M. J., Catlow, C. R. A. & Lewis, D. W. Modeling Aqueous Silica Chemistry in Alkali Media. J. Phys. Chem. C 111, 18155 – 18158 (2007).

[12]   Trinh, T. T., Jansen, A. P. J. & van Santen, R. A. Mechanism of Oligomerization Reactions of Silica. J. Phys. Chem. B 110, 23099 – 23106 (2006).

[13]   Pereira, J. C. G., Catlow, C. R. A. & Priceb, G. D. Silica Condensation Reaction: An Ab Initio Study. Chem. Commun. 6, 2 – 4 (2012). URL http://pubs.rsc.org.

[14]   Pereira, J. C. G., Catlow, C. R. A. & Price, G. D. Ab Initio Studies of Silica-Based Clusters. Part II. Structures and Energies of Complex Clusters. Microporous Mater. 103, 3268 – 3284 (1999).

[15]   Thuat, T. T. A Computational Study of Silicate Oligomerization Reactions. Ph.D. Thesis, Eindhoven University of Technology (2009).

[16]   Becke, A. D. Density-Functional Thermochemistry. III. The Role of Exact Exchange. J. Chem. Phys (1993).

[17]   P. J. Stephens, C. F. C., F. J. Devlin & Frisch, M. J. Ab Initio Calculation of Vibrational Absorption and Circular Dichroism Spectra using Density Functional Force Fields. J. Chem. Phys (1998).

[18]   Tossell, J. A. Theoretical Study on the Dimerization of Si(OH)₄ in Aqueous Solution and its Dependence on Temperature and Dielectric Constant. Geochim. Cosmochim. Acta 69, 283 – 291 (1981).

[19]   Nangia, S. & Garrison, B. J. Reaction Rates and Dissolution Mechanisms of Quartz as a Function of pH. J. Phys. Chem. A 112, 2027–2033 (2008).

[20] M.W.Schmidt, J. S. M. J. S. N. K. S. T. M. J., K.K.Baldridge. General Atomic and Molecular Electronic Structure System. J. Comput. Chem 14, 1347-1363 (1993).

[21] Brett M. Bode, M. S. G. Macmolplt: A Graphical User Interface for Gamess. Journal of Molecular Graphics and Modeling 16, 133-138 (1998).

AUTHOR INDEX

SUBJECT INDEX